Is God a Scientist?

Also by Robert Crawford

CAN WE EVER KILL? (*second enlarged edition*)

JOURNEY INTO APARTHEID

MAKING SENSE OF THE STUDY OF RELIGION

A PORTRAIT OF THE ULSTER PROTESTANTS

THE SAGA OF GOD INCARNATE (*second enlarged edition*)

THE GOD/MAN/WORLD TRIANGLE

WHAT IS RELIGION? (*translated into Greek and Portuguese*)

Is God a Scientist?

A Dialogue between Science and Religion

Robert Crawford

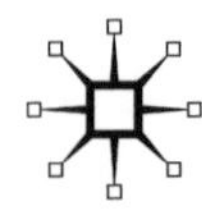

First published 2004 by
PALGRAVE MACMILLAN
Houndmills, Basingstoke, Hampshire RG21 6XS and
175 Fifth Avenue, New York, N. Y. 10010
Companies and representatives throughout the world

PALGRAVE MACMILLAN is the global academic imprint of the Palgrave Macmillan division of St. Martin's Press, LLC and of Palgrave Macmillan Ltd. Macmillan® is a registered trademark in the United States, United Kingdom and other countries. Palgrave is a registered trademark in the European Union and other countries.

ISBN 1–4039–1688–8

This book is printed on paper suitable for recycling and made from fully managed and sustained forest sources.

A catalogue record for this book is available from the British Library.

Library of Congress Cataloging-in-Publication Data
Crawford, Robert G. (Robert George), 1927–
Is God a scientist? : a dialogue between science and religion / Robert Crawford.
p. cm.
Includes bibliographical references and index.
ISBN 1–4039–1688–8
1. Religion and science. I. Title.

BL240.3.C73 2004
201′.65–dc22 2004049124

10 9 8 7 6 5 4 3 2 1
13 12 11 10 09 08 07 06 05 04

Printed and bound in Great Britain by
Antony Rowe Ltd, Chippenham and Eastbourne

To the memory of Ivor Lewis and James Haire

Contents

Introduction 1

1 The Method of Science 7

2 Religion and Scientific Method 23

3 God Talk 37

4 The Experiment Begins 57

5 The Subjects of the Experiment 75

6 The Test 93

7 The Action of the Divine Scientist 111

8 The End of the Experiment 133

9 Objections to the Model of the Divine Scientist 155

Notes 165

Index 173

Introduction

Many religious people think that science has nothing to do with their faith and they try to forget how science views the world and us. But if God has created the world then what science discovers about it may help us to understand him. In the past scientists have said that God has two books, one of scripture and one of nature. If this is so then religion and science should complement one another as Copernicus, Galileo and Newton believed.

Those who package religion into one area of life and science into another usually do so because they believe there is a conflict between the two disciplines. There has been in the past, as the debates in the nineteenth century show, but religion did try to find a way of reconciliation and it is continued by some scientists and theologians today.

But is science not based on fact while religion relies on faith? This is true to some extent but science has a number of beliefs which cannot be proved. It proceeds on the basis that there is order and regularity and that we can discover how the world behaves. That is all it can do since we cannot know what the world is like in itself but how it behaves. Religion also proceeds on the basis that God exists (Heb. 11:6) and that he is the rewarder of those that seek him. It is significant to note that the classical arguments for the existence of God were put forward by those who had already exercised faith: Anselm, Aquinas, Descartes. It was faith seeking understanding.

It is clear that religion is interested in the order and beauty of the world and the meaning of it all, but personal problems about sin, suffering and forgiveness are the main area of focus. Science in general, however, is more impersonal and often deals with objects – apart from the social sciences, which deal with people and attempts to understand them by participation in their lives. Some offer the personal counselling which was once reserved for religion. What we will point out is that even in physics there is the involvement of the observer as she seeks to understand quantum mechanics.

But can religion be content with the inner aspects of our lives and neglect what science is saying about the world? Is God not involved in nature as the Bible insists? Indeed the psalmist not only wonders at the beauty and the order of the world but the man that God has created (Psalm 8). He would surely have been interested in the way scientists regard not only the world today but also, with their stress on our genetic inheritance, the nature of mankind. The psalmist thought that the world was designed by God and man was made in his image. What would he say to those scientists who deny it today? We do not divide life into different compartments; what we do in one influences another. Why then seek to totally separate religion and science? Both disciplines make cognitive claims and they need to be able to defend them.

A useful distinction, however, is that religion deals with the *why* question and science with the *how*. Science tells how the universe came into existence whereas religion explains why it exists. Many scientists interested in religion take up this position, but the two questions intersect because when the scientists tell us that the universe originated from the Big Bang the question arises how does God as creator fit into this scenario? Perhaps that is why Einstein said: 'in every true searcher of nature there is a kind of religious reverence. ... science without religion is lame, religion without science is blind'.[1]

Religion cannot be based on science but its views should have some consonance with the scientific understanding of the world and mankind otherwise it fails to communicate with a scientific society. Conversely, if science seeks to proceed without any acknowledgement of the values and purpose which religion imparts to life, its methods may become inhumane and its goals meaningless.

Today many theologians and scientists are seeing parallels between the disciplines rather than differences. Science is not viewed as totally objective and religion subjective. It used to be thought that the scientist was a detached observer but in relativity the basic measurements, such as the mass, velocity and length of an object depend on the frame of reference of the observer. There is the personal participation of the knower in all knowledge. Assumptions enter into both disciplines together with creative imagination and personal judgement. But in religion there is more personal involvement since the goal is the changing of the person.

When a scientist selects, reports, and interprets data, he has a theory in mind. We may say that the data of religion are scripture, religious experience and rituals, and there are various views about their place and importance. The language of both science and religion include analogies, models, metaphors, similes, myth, and so on. Science is done in a community and theologians cannot proceed without taking into account the community of believers and the history of religious experience.

The scientist tests his theories by experiment and it is clear that the possibility of testing religious beliefs is more limited than in science. But faith is an experiment leading to the experience of the love of God and we are called upon not to rely on the faith of others but personally to commit ourselves. One major test resulting from such an experience is the love of God and the neighbour (1 John 4:20) Faith will show itself in loving behaviour not only to those who love us but others who differ from us in various ways.

Religion continues to interest scientists, with the media bringing together thinkers from both disciplines to discuss the existence of God, the role of faith, the implication of genetics and so on. Scientific surveys continue to probe the extent of belief and the dangers or benefits of it. While some psychologists continue to argue that it is a projection of the father-image or of our wishes for a better life, sociologists see it as holding societies together and providing values. Many who helped the advance of science were religious believers and it can be contended that their view of the world as orderly and being designed encouraged others to investigate. We will oppose the view that religion is in conflict with science whose role is to dispel religious superstitions and mysteries and gives an ultimate explanation of everything.

The success of science has given it prestige. Because of its labour we know that this vast universe has millions of galaxies and laws which can be described mathematically. We now know much more about matter and the genetic structure of our bodies and brains and can make plans for the future that previous centuries could only dream about. But there is the other side of science which has ushered in an apocalyptic nightmare: atomic and hydrogen bombs, deadly viruses, pollution, and so on. It is clear that religion with its values needs to be involved in what science can and cannot do. We acknowledge the prestige of science but will argue that its explanations can often be incomplete and require the help of religion. Science and religion complement one another in understanding what we are like and the world in which we live.

But since we live in a scientific world it would seem appropriate to find a new model of God and in this book it is argued that there are grounds for portraying him as the cosmic scientist. If God is a scientist then the laboratory in which he works is the world and his intention is to create a humanity which will obey his laws and live up to the values which he embraces. But it is essential to show the differences between him and the human scientist in the conduct of experiments. With the human scientist it is sometimes the case that the main concern of an experiment is success rather than the welfare of those involved. By way of contrast we will contend that the divine experimenter shows love and compassion for the subjects of the experiment who are in no way treated as objects. In particular they possess the freedom to obey or disobey his laws.

Some argue that science and religion are so different that it is impossible to see any similarity between their method of understanding and the language they use. We disagree and the opening chapters will try to show that even though religion is concerned with God and science with the world and mankind, the two disciplines overlap. What science says will also have implications for our view of God. Hence Chapters 1 to 3 are devoted to the methods employed by the scientist and the similarities and difference between the human and divine way of working.

Chapter 4 deals with the beginning of the experiment: the creation of the world and mankind and takes into account what physics and evolution are saying and argues that God has given both the world and us freedom to develop. But in Chapter 5 there are signs in scripture that the experiment has gone wrong. God, however, is interactive in the experiment and makes new plans: covenant with Noah, Abraham, Moses, and the coming of Christ and the rise of such prophets as Muhammad and Guru Nanak. This involves suffering both for God and us but it is aimed at the triumph over evil. It is a major problem for religion and is discussed in Chapter 6.

How God might act in the experiment is discussed in Chapter 7, taking into account what science says about the world and mankind. Chapter 8 discusses the end of the experiment and tries to show that the religious picture, which does not need to be taken literally, is much more optimistic than that of science.

In a previous book I drew on six world religions in the dialogue with science but in this one I have concentrated on the Judaeo-Christian tradition and only mentioned other religions when appropriate. It is hoped that this will make the discussion more concentrated.

God does not have a gender but it is impossible to write about the concept without using gender-specific terminology. I use the masculine pronoun with reference to the deity but recognise in Chapter 3 that female models exist both in scripture and contemporary writing. With regard to ourselves I have tried to alternate between 'she' and 'he' but 'man' is used to include male and female.

I would record my appreciation to my wife for her reading of the manuscripts and proofs and to Mr John M. Smith,

Senior Book Editor at Palgrave Macmillan, for his detailed work in editing the book. Throughout I have tried to give references to the various writers whom I have consulted but if anyone has been omitted I would apologise in advance.

ROBERT CRAWFORD

1
The Method of Science

In *The Sign of Four*, at the scene of a crime, Holmes says to Watson, 'I want you not only to look but also to observe what has happened here.' Observation needs to be careful and focused if clues are to be found as to who did it and how. Charles Darwin (1809–82) realised the value of careful observation and noted what others who were more learned missed. It led to the theory of evolution, which he thought explained the facts. The result was a discovery about our origins and development which disturbed and shocked the world. We have all seen apples falling from trees but few of us are like Isaac Newton (1642–1727), who, if the story is accurate, saw an apple fall and thought of the law of gravity.

Scientists and philosophers make use of induction, that is, having observed a large number of occurrences under different conditions, they establish a general law which predicts what is likely to happen in the future. If for a long time I have noted that all swans are white I conclude that this is the case, but my conclusion may be wrong, for a black swan may turn up somewhere. Bertrand Russell tells the story of the inductive turkey. The turkey noted that he was always fed daily at nine a.m. and he made these observations under a wide variety of circumstances; hence he concluded that he was always fed at that time. But he reckoned without Christmas Eve when instead of being fed he had his throat cut!

Other philosophical objections to the method of induction were made by David Hume (1711–76), but observation itself presents a problem, for observers may not see the same thing and the scientist will observe more than the amateur. The expert brings past experience, knowledge and expectations to her observation and often the theory that emerges is laden with these. The kind of questions asked will determine the answers with theories involving novel concepts and hypotheses not found in the data, that is, referring to entities and relationships not directly observable. As we will see when physicists study the sub-atomic world, observation can alter the object observed, so the participant observer is part of an interactive system.[1]

Induction has been of great value, as shown by the work of the famous scientists. Galileo carried out many experiments with balls on inclined planes and generalised about the behaviour of bodies in motion. Deduction from the general to the particular is important and considered more certain. Newton, having established that there is a force directed towards the sun, showed deductively that planets sweep out equal areas in equal times as they go round the sun. Simple examples of deduction are:

> All men are mortal, Socrates was a man, therefore Socrates was mortal.
>
> Water heated to 100 degrees centigrade will boil, that sample of water has been heated to 100 degrees centigrade, therefore it will boil.

In such arguments we have premises and a conclusion but the former must be true for the conclusion to be true. Many scientific theories cannot be directly tested but deductive reasoning can deduce from the theories consequences that can. Thus we might speak of the 'top-down' approach of deduction and 'bottom-up' approach of induction. Nature, however, has many surprises in store for the scientist; for example, in quantum physics the electron which is one of the constituents of an atom is uncertain in its behaviour. If you know where it is you cannot know what it is doing and if you know what it is doing you cannot know where it is![2]

In contrast to this behaviour at the micro-level, nature at the macro-level is predictable. Newton was concerned with this level and one discovery led to another. When he saw the apple falling to the ground he realised that matter attracts matter with gravitational forces working at great distances between bodies in space. The same force that pulled objects to earth kept the planets in their orbits: the law of universal gravitation. But observation needed to be supplemented by imagination and mathematical ability, with Newton showing more superiority in these areas than his rival Robert Hooke, who tried to claim credit for the ideas about gravitation. But Newton's theories were not proved until 1798 when Henry Cavendish did a laboratory experiment that verified the law of gravitation.

The laws of Newton still hold, that is, a body at rest stays at rest and a body in motion remains in motion at a constant speed and direction as long as outside forces do not interfere. When an outside force changes the speed and direction of a moving object, the rate of change is proportional to the amount of force exerted; and for every action there is an equal and opposite reaction. The planets move in a pattern but when an exception to it was noted (Uranus), it caused a lot of debate. Was it due to the planet being furthest from the sun or was there another planet, Neptune, whose pull on Uranus made it behave incorrectly? It was the latter view that proved ultimately to be correct.

Problem-solving

Science often starts with a problem that needs to be solved. Gregor Mendel, an Austrian monk, lived in the 1800s and was puzzled by the apparently haphazard distribution of presumably inherited characteristics in plants. He proceeded to construct a hypothesis concerning the laws of inheritance so that, seen in this context, the pattern of characteristics found in each new generation is no longer a puzzle but predictable. Darwin was unaware of Mendel's work but its discovery at the end of the century led to the modern study of genetics.

Karl Popper saw science as problem solving but insisted that in the past scientists have looked for verification of their theories instead of falsification. Theories are conjectures or guesses which are then tested by observation and experiment. Progress is made by trial and error and the fittest survives. We can never say a theory is true but the best we have at the moment. It must be possible to falsify theories but Freudian psychoanalysis, Adlerian ideas, and some versions of Marx's theory of history could not be falsified. Adler's ideas are consistent with any kind of behaviour since they rely on a sense of inferiority. If a man leaps into a river to save a child he combats such inferiority but if he does not he is showing his strength of will to remain unperturbed while the child drowns. This is probably a caricature of Adler but his explanation of behaviour is so vague that it cannot stand up to the clarification demanded by falsification, and the same is true of the theories of Freud and Marx.

Popper believed that the progress of science lay in conjecture and refutation and, since the latter continues to apply to theories, we can never know if they are true. All we have is a more accurate model; we can only continue to seek the truth. Absolute truth is not attainable.[3] Popper used deduction but he fails to disclose the positive content of science in any detail. We are left with negative results rather than positive. The question might also be raised about his work: how do we arrive at the meaning of falsification, is there not a need to pay more attention to the context? Against Popper, it was argued that if future observations contradicted a theory they may also confirm one which has previously been falsified and it is difficult to devise an experiment which will dispose of a theory in its entirety. Perhaps it was only one part of a theory that presented difficulties and it might still be retained by the use of auxiliary hypotheses.

Newton's theory was extremely resistant to falsification but Einstein was able to account for phenomena which falsified it. Bold conjectures involving risks led to an advance as in Einstein's risky prediction that light rays should bend in strong gravitational fields. Theories can never be established as true or probably true so the aim of science is to falsify them and replace them by better ones which demonstrate a greater ability to withstand tests.[4]

Revolutions in science

Thomas Kuhn (1922–96) paid more attention than Popper to the context of the development of science and drew attention to paradigms which are frameworks or procedures or ideas that inform scientists how to work and what to believe. Scientists work in communities with shared methods, ideals and aims. Most do not question these but proceed in the light of them, ignoring anomalies, and seeking solutions to reinforce the paradigm. But anomalies may accumulate and cause a revolution or a major shift in science. The old must give way to the new with no part of the old being retained. Reason is involved but a paradigm could be compared with a conversion, a new way of seeing things. Such a major shift happened when quantum physics and relativity replaced classical physics. Earlier Kuhn had thought that paradigms are incommensurable (cannot be directly compared with each other), but later he acknowledged that there could be agreement on observation statements despite rival paradigms and it might lead to some sort of compromise.[5]

It is generally agreed that Kuhn put too much stress on sudden revolution and lack of continuity and neglected smaller factors. His work raised the question: do theories emerge from experimental evidence or from cultural factors? The latter figured prominently in his views. Kuhn thought that scientists can never understand the real world and in opposition to Popper held that falsification is no more possible than verification since there are no absolute standards to measure any paradigm. There are different paradigms for different cultures so they arise in communities and may not be determined by experimental data or by normal research.

Ian Barbour summarises Kuhn, taking into account his later modifications: all data are paradigm-dependent but there is data on which adherents of rival paradigms can agree. Paradigms are resistant to falsification by data but they do cumulatively affect the acceptability of a paradigm. There are no rules for paradigm choice but there are shared criteria for judgement in evaluating paradigms. This saves Kuhn from subjectivism. But it seems clear that accounts of science that were inductivist or falsificationist did not stand up to the historical evidence. Theories arose in an historical context with the sociological characteristics of scientific communities playing a role. Sociologists were keen to build on Kuhn's foundations and he is considered to be one of the roots of postmodernism. Kuhn opposed the rationality of Popper, arguing that paradigm choice can never be unequivocally settled by logic and experiment alone. There is a need to take account of intuition, ideological stance, aesthetic taste, in brief a non-rational component.[6]

Imre Lakatos (1922–74) tried to save the content of science from the full implications of Popper's falsification by stressing that scientists did not always abandon their theories when the results were negative.[7] We must wait until there are better theories to replace them, for scientific programmes have a hard core protected from falsification by auxiliary hypotheses. The hard core of Copernicus was that the earth goes round the sun and spins on its axis once a day and with Newton there were the laws of motion and law of gravitational attraction. Both were augmented by modifying or replacing auxiliary hypotheses which supplemented the hard core. Modification of the auxiliary hypotheses can lead to some novel discoveries but Lakatos did not offer a clear-cut criterion for rejecting any coherent research programmes or for choosing between rival programmes. However, he paid attention to the actual practice of scientists and saw the need to test and improve the auxiliary hypotheses.

Objective or subjective?

Three approaches to scientific knowledge have been noted:

(1) Subjective: science is a set of beliefs based on sensory experience or introspection and reasoning on the part of the individual scientist.

(2) Consensus: a theory must gain the assent of the community and its members who are trained scientists. The consensus is relative to a particular community and meets its needs.
(3) Objectivist: science is not a set of beliefs either individual or collective because theories are autonomous and independent of (1) or (2).

A.F. Chalmers contends for the objectivist position because the production and appraisal of scientific knowledge is a complex social activity which undermines both the subjectivist view and the consensus. The subjectivism of the individual is curbed by tests and she needs the cooperation of others. The consensus view is inadequate because it believes that scientific theory or practice is the possession of a social group. But for the objectivist, theories and practice are primary and they are there whether or not a particular group supports them. Science is international: Galileo in Italy, Newton in England, Einstein in Germany and America, and so on.[8]

Popper would appear to agree with Chalmers since he argues that a theory is true or closer to the truth whether an individual or group think so or not. But he speaks about three worlds: the physical, the mental (subjective thought processes), and the world of ideas. Mind is the link between the first world and the third so there is a subjective element but materialists reject such a third world where theories or ideas exist autonomously. Theories for them form part of an objective scientific practice and it modifies and produces new theories, which removes the subjective element. Lakatos argued that there is a single timeless correct scientific method and we need to discover it but, as we saw, he offers no theory of how this is to be done. Chalmers thinks the best we can do is to investigate science as a practice.[9]

Scientific terms must be precise: Newton's theory of mass has a precise meaning because it plays a role in a structured theory, and contrasts with a vague term like democracy. Concepts are defined in terms of the meaning of other concepts. Theories are seen as organised structures of some kind and in that structure they acquire a precise meaning. The more specific a theory is the better; for example, the planets move around the sun under the influence of a central force which varies exactly inversely as the square of the distance from the centre. It has a precision which contrasts with general statements. There is also the principle of simplicity: the simplest of two or more explanations which are equally convincing is more likely to be true. A belief in the simplicity of nature has been effective in the work of Galileo, Kepler, Newton and Einstein.[10]

Michael Polanyi (1890–1976), however, contended that knowledge is both objective and subjective so would oppose Chalmers. Quantum mechanics has shown the influence of the observer on the experiment and in general the scientist makes a personal contribution of commitment, skill and judgement. He stressed the part played by this interpretive skill in selecting observations and making judgements about experience: hence the subjective element cannot be avoided.[11]

The scientist, as we have seen, relies on observation, gathers data, experiments, and then produces a theory but he may begin with an idea, a guess or a problem. There is no precise answer to how the scientist gets the idea in the first place. Some think it is due to creative imagination or inspiration but to get a result there must be a combination of reason and experience. Stephen Hawking spoke of a leap of faith or inspired guesswork regarding the best theory or model of the universe or which equation was to be used.[12] Before Einstein the stress was on the empirical: experiment, observation, use of apparatus; but he engaged in 'thought experiments' and arrived at conclusions some of which took many years to confirm. He envisaged atomic explosions and rockets to the moon before their actuality. Intuition is involved as in the case of Darwin or due to an accident: the discovery of X-ray plates. The work of sheer genius defies logical explanation.[13]

Testing

The testing of theories is difficult and relevant factors can be omitted. Such omission caused Heinrich Hertz in 1888 to fail when he performed the electrical experiment to produce and detect radio waves for the first time. He was testing Maxwell's electromagnetic theory to produce the radio waves predicted by that theory but when he measured the velocity of radio waves he failed to take into account the dimensions of the laboratory in which he worked. Radio waves emitted from the apparatus were reflected from the walls of his laboratory on to the apparatus and they interfered with it. Hertz thought the conditions of the laboratory did not matter because he had a theory in mind and sought only facts relevant to it.

Testing will involve the judgement of other scientists and even publication does not guarantee success since it may be thought that the result is not fruitful in suggesting other new problems and development of techniques. If the research establishes a fact about nature (some philosophers dispute whether anything can be called a 'fact'), it enters the text-books and may lead to a new law since it gives a unified description of a whole range of phenomena.[14]

Barbour contends that there are four criteria for assessing scientific theories in normal research. There must be agreement with data, predictive success, coherence and interconnection with other theories. Support for a theory is strong if there is a variety of evidence and it is comprehensive. Valuation can be based on the measure of fertility in furthering an ongoing research programme and in generating new hypotheses and suggesting new experiments. The difficulty about the data is that it is not theory-free and many theories postulate unobservable entities. His conclusion is that science does not lead to certainty because theories change in time and are modified but science does have reliable procedures for testing and evaluating theories by complex criteria.[15]

Limitations

Some scientists seek the behaviour of things only in terms of their component parts, that is, reductionism. But now there is a more holistic approach following criticisms by Paul Feyerabend, James Lovelock and others. Quantum theory has shown that the relationships between entities are important, with the total system playing a vital part, and attention has been paid to the emergence of higher-order behaviour from matter in non-reducible ways. Science is limited in not being able to assess work in other areas of importance such as music, poetry and art, for a different kind of evidence is required. There is also 'why' a universe exists and while science is good at the 'how' it has difficulty with the 'why'. But we want to know why we are here and why there is something rather than nothing.

The more science probes into the material of the universe the more surprising it appears. Scientists debate about the nature of matter since so many particles have been discovered that it can be compared to peeling an onion. Quarks and leptons may be the ultimate constituents but there may still be more levels or they never come to an end. Black holes occur when stars collapse down to a point which does not reflect any light but they are known to be there from the way they swallow up anything that passes too close to them. There is also dark matter in the universe which holds the galaxies together and there is at least ten times more of it than can be observed.

Science has been driven to deal in paradoxes. It works with two kinds of time: the one we know of past, present and future, and the static time of physics with no special moment. Which one is true? Then there is the Heisenberg Uncertainty Principle which means that we have to live with probabilities not certainties. Are we to regard the behaviour of a particle as a wave or a particle? What is an electron? We know how it behaves but not what it is. The question asked is important. If we ask how light or an electron interacts we use the language of particles but if wanting to know where the light interacts we use the language of waves!

Quantum theory insists that we cannot know the locality and velocity of particles at the same time, uncertainty relates to the measurement. Since Einstein believed in determinism he tried to disprove that uncertainty was inherent in nature. At a conference in Brussels, he asked Bohr to visualise an enclosed box containing a radioactive element and an alarm clock. The alarm, set for a precise time, operated a camera-like shutter in the box that released a certain amount of radiation. Imagine that the box is weighed before and after the release of the radiation. E = mc squared, so the mass is equivalent to energy; hence the loss in weight (mass) can be translated into energy. If one can determine both the energy of released radiation and when it occurred there was a direct contradiction of the uncertainty principle.

Bohr replied that Einstein's own theory refuted the 'thought' experiment, for general relativity states that a clock not influenced by gravity runs faster than one in a gravitational field. This was confirmed later by testing atomic clocks in airplanes and satellites. When the the alarm clock was exposed to the varying effects of the gravitational field and the exact time the shutter opened and released was noted the energy would be uncertain. Later Einstein admitted that quantum mechanics does contain part of the ultimate truth.[16]

The general position appears to be that quantum is defined by the act of observation. We need to ask questions before we observe the electron, which is neither a wave nor a particle at this point. It is unreal, in some kind of limbo, so we create the truth, making an imaginative world. David Bohm disagreed and insisted on the implicate order. Particles are particles not just when they are being observed but their behaviour is determined by a pilot wave and measurement alters the wave. One particle can influence another instantaneously across distances and there is a deep implicate order under the explicate order of the world. It is due to an infinite number of fluctuating pilot waves. Any attempt to measure one's thought changes it just as the measurement of an electron alters its course. But he went on to admit that reality is basically unknowable.

Limits on the ability to frame or test a theory of everything spring from Kurt Gödel's incompleteness theorem in mathematical logic which implies that even within mathematics there are doubts. Certain questions occur that cannot be proved or disproved in a mathematical system on the basis of axioms or assumptions within the system. It is uncertain that these axioms will not give rise to contradiction.

Scientific laws predict what to expect and must be consistent with one another. Reason is the tool working out the mathematical formulae which describe nature, but it is recognised that nature has an element of mystery. There is uncertainty about the laws since further observations may call for a change in them. Experiments are never complete and attention to small details can result in change, hence the scientist must leave room for doubt. Logic can have its problems and science changes its views as knowledge develops. The steady state theory has been replaced by the Big Bang and now there is talk about a number of Big Bangs. There are questions that cannot be answered: How did the laws of the universe come into being, what is the purpose of it all, why should we hold certain values and not others? Science is limited to the explanation of the internal affairs of the world.

There is uncertainty when new circumstances arise, such as knowing what is within black holes or in the electron or laws that operate at the speed of light. Do the laws which we know always operate? Did they apply at the start of the universe when there were extremes of density and temperature? A theory is useful, not certain, for another one can supplant it: Newton's theory of gravity, attraction of objects, was replaced by Einstein's theory of gravity, distortions in space. Matter for Newton was invariable but Einstein saw that it was variable and dependent on motion. Causality was deterministic with Newton but probabilistic with Heisenberg. Einstein's theory of relativity opposed Newton. Einstein argued that light close to the sun would be bent by the curvature of spacetime and during the total eclipse of 1919 he was proved right. But surprises were in store for Einstein. Georges Lemaitre (1894–1966), a priest, pointed out that if the galaxies are moving apart, they must have been closer together in the past so at one time all the matter had been concentrated into a single point with a high density. The theory heralded the idea of the Big Bang origination of the universe.

While science deals with facts there are various interpretations. Thus there is disagreement between the quantum experts: Is quantum indeterministic (Bohr) or is it deterministic (Bohm)? Philosophers of science differ, as we have seen, with Popper teaching that science stops at the first sign of falsification but Lakatos arguing that theories can be accepted even if they do not perfectly fit observation. Taking this into account, Michael Polanyi argued that science was pursued by persons, based upon personal judgement and commitment to a point of view. But restraint is placed upon individuals by the scientific community which evaluated the theories. Thus science was like a Faith community open to the testing of its belief (1 Thess. 5:21). But subjective criteria such as elegance and beauty determine some theories since they cannot be empirically verified.

J.B.S. Haldane (1860–1936) was sure that the universe is not only stranger than we imagine, it's stranger than we can imagine. Einstein contended that we cannot have information at speeds faster than light and chaos theory confirms that many phenomena are impossible to predict. Stephen Hawking, speculating about matter and other universes, notes that quantum uncertainty causes matter and energy, space and time, to flicker between different states. He contended that spacetime fluctuations may cause wormholes which could link one area of spacetime with a distant one or two 'baby universes'. He pursues a Theory of Everything but in 1994 he admitted that science may never find one.[17]

Just like religion, science is now dealing with invisible entities. The believer speaks of God, faith, grace, atonement, an unseen world; while the scientist refers to atoms, electrons, nuclei, viruses, hormones, genes, which are difficult to detect and are usually known by their effects. The meaning of mass, force, entropy, is not exhausted by observable properties so pure empiricism is an insufficient basis for science. Scientists appear content with verisimilitude, approximate truth, and explanatory power. Neither the models they use, examined in Chapter 3, or the mathematics, are the final description of the world. Success has often depended upon the invention of instruments such as the microscope and telescope. The first enables the scientist to see organisms too small to be detected by the naked eye and the second enabled Galileo (1564–1642) to chart the mountains and craters of the moon. Today electron microscopes can magnify an object up to two million times its actual size. It is accepted that even if electrons, quarks and neutrons in the world of quantum cannot be observed directly they do exist. Critical realism accepts this indirectness and calls for models and analogies to help us to understand behaviour.

Some sociologists, influenced by postmodernism, challenge science and contend that theology and science are different ways of constructing the world. The non-realist also questions whether science gives us an objective and real world and, fastening on Kuhn's point about the scientific communities, asks if the theories were manufactured by them. They oppose the view that experimental decisions are governed by strict rules and say that they are due to negotiations between scientists.

Science does not discover the world, it makes it. Basically it gives us a social construction for it is the scientist, influenced by the culture and grants that are available, who decides what facts are significant. Kuhn said that the scientific community exchanges one paradigm for another partly on non-scientific grounds. Sociologists even go so far as to ask the question: Was Darwin's theory formed because the needs of his society demanded competition and the survival of the fittest rather than a more cooperative model which might have fitted the 'facts' equally well?

Critics of this viewpoint call it an extreme and say that sociology is always trying to explain matters in terms of social processes: that is its job. It does not answer the question how the scientific community can accept a new truth when it does not appear in the interests of society. Sociology can criticise knowledge at the periphery of science but not the hard core.[18]

In summary, the scientific method consists of data, measurements and observations and testing of ideas by experiment; theories which use the data to explain how things work; and shaping principles. The principles are extra-scientific values and assumptions, background beliefs and commitments which influence data gathering and theory forming. Physical science proceeds from hypothesis to theory and then to law, and these are described mathematically and predictions are made. Even in quantum physics large events are predictable but living systems are more complicated, which means dealing with probability rather than certainty.

The order of the world is assumed. Theories are not absolutely certain. For example, is the DNA determined by RNA or vice versa?[19] There are also problems in connection with research programmes. What theories will I use and what data will I exclude, how can I escape being influenced by funding, and so on? Yet the scientist tries to maintain objectivity and the values of honesty, faith in himself and his research and depends on the scientific community scrutinising his information and contributing to it.

Despite what positivism says, science cannot explain everything; that is not science, but scientism. It is the latter that mounts assaults on religion, as we will see in subsequent chapters.

2
Religion and Scientific Method

At first sight there seems little hope of seeing similarities between science and religion not only because the vocabulary and content are so different but in the light of past conflict. However, it should not be forgotten that Judaism, Christianity and Islam encouraged the development of science by denying, unlike the ancient Greeks and others, that nature was divine; it could then be explored experimentally. With the belief in a rational and orderly world science made great strides and James Clerk Maxwell felt justified in placing above the door of the Cavendish Laboratory in Cambridge the text from Psalm 111, 'The works of the Lord are great, sought out of all them that have pleasure therein'.[1]

But, as we saw in the last chapter, science is regarded as rational based on evidence and superior to religion which relies on faith. It is always open to new ideas and criticism whereas it is said religion is dogmatic and denies empirical testing. But we also noted that the data of science is theory-laden, and not free from value judgements. Science is a cultural and social phenomenon with scientists participating in what they observe. The postmodern approach points to conflicting and competing paradigm theories, research programmes and traditions and tells us to trust more in what is going on in local practical contexts.[2]

Of course, there are differences between scientific and religious methodology, because the object studied is different: religion focusing on God and science on the world. But if

God is creator of the world then we hope in this chapter to see him reflected in his handiwork as we compare and contrast the methods of the two disciplines.

Science seems more believable because its object can be observed. At least this was the traditional view; but now, as we mentioned, science often has to deal with invisible entities: quarks are a level in the structure of matter and are stuck together by gluons but no one has ever seen an isolated quark. Do they exist physically or only mathematically? Dark matter is present in the universe but is not observable. Science relies not only on the empirical but also on the non-empirical, that is, economy, elegance, and naturalness of theories, so theory and experiment are interwoven.

Data of religion

The data of religion are revelation, scripture, religious experience, worship and ritual, and these interact. Religious experience can take the form of an experience of forgiveness or the feeling of the numinous or awe. Events can create it or a response to the moral demands of conscience or becoming aware of a spiritual presence in the beauty of nature.

Experience of God may occur in a religious community or elsewhere and lead to a paradigm shift, as was the case with Martin Luther in the sixteenth century. Though he did not want to divide the church he was excommunicated and Protestantism resulted. But Catholicism, while it carried out reforms, persisted with the usual paradigm. Political and economic factors played a part in the dispute but for Luther it was the interpretation of scripture that was crucial. Buddhism, born in the Hindu context, became a new paradigm in which the soul was denied and self-help rather than the grace of God was stressed. The rise of Sikhism could also be considered a major shift, though some have argued that the Faith is a mixture of Islam and Hinduism. But the view cannot be maintained since there was something new and original, present in their founder, Guru Nanak. Muhammad also initiated a revolution in disposing of the Arabic gods in favour of monotheism. But there is continuity between the old and the new paradigms in some of these cases: for example, belief in the idea of a soul continued in Sikhism and Luther's protest in Christianity with its stress on justification by faith did not dispense with such doctrines as the divinity of Christ or the Trinity. Hence it is often better to speak of evolution rather than revolution.

As in science, religion holds on to a current paradigm until matters reach a crisis. A hard core of ideas is protected from falsification by auxiliary hypotheses. Nancey Murphy argues that the way forward in religion is to study the practices of the Christian community, including worship and scripture, in a continuing research programme which resembles the ideas of Chalmers in science. Ian Barbour proposes that process theology can be the theological programme in which the hard core of the Christian tradition is creative love, revealed in Christ. Divine omnipotence is treated as an auxiliary hypothesis or peripheral belief, modified to allow for human freedom, evil and suffering, and evolutionary history. But theologians in the various faiths differ about how we are to proceed. Some stress scripture and give it a supreme place (Islam, Sikhism, Protestantism). Others emphasise liturgy or morality or the community or... . Traditionally in Christianity scripture, tradition, faith and reason have been to the fore with different degrees of emphasis. This fits in with science which demands intelligibility, coherence, testability, and objectivity.[3]

We noticed inspiration and creative imagination in science, and Darwin said that he could recall the moment in his carriage when he grasped the significance of evolution. But there are many other examples of this kind of insight in the history of science. For example, there was the case of carbon atoms which can form bonds with other carbon atoms. They form a ring of six atoms, the benzene ring. A. Kekulé puzzled over this and had a kind of vision, a waking dream, while riding on a horse-drawn London bus in 1865. He saw chains of carbon atoms dancing around and then one of the chains looped round and grabbed hold of its other end to make a circle. Thus he concluded that carbon atoms in a benzene ring are arranged in a circle and each carbon atom uses three of its bonds to keep the circle closed, forming a double bond on one side and a single bond on the other. It leaves each of the six carbon atoms with just one bond free, sticking out from the circle, with which to hold on to a hydrogen atom.[4]

In the case of religion there are visions, dreams, and various kinds of inspiration. These form part of the content of revelation which is a spiritually uplifting sense of God conveying a meaning beyond itself. The experience can come through dreams, as in the case of Joseph, or in the form of a warning (Mt 2:12, 13) or in a vision, as when John the Baptist recognises the holiness and spirituality of Jesus and sees a dove (the Spirit) descending upon him, or the visit of an archangel as with Muhammad. It can also occur in the wider context of nature: a burning bush conveying a message to Moses. Revelation is of major importance for religions, since their founders (Hinduism does not have one) testified to have received it from God.

But we do not have direct evidence that these experiences took place: they are mediated to us through the scriptures or worship of the church. Such indirect evidence can be queried but science at times does not have direct evidence for invisible entities like electrons and we need to remember that God is not a physical object. The Bible which puts forward the evidence for his existence is now regarded by most Christians not as revealed propositions but a witness to Christ. Muslims believe that the Qur'an was dictated to Muhammad but it has not been subjected to the historical criticism applied to the Hebrew and Christian scripture which we will see later. Theologians are inclined to follow Karl Barth who did not equate the Bible with the Word of God but insisted that we find the Word (Christ) amid the words. However, he isolated theology from science by contending that the former had its own character and rationality.

Revelation

Many books have been written on revelation, but most Christian writers believe that revelation takes place in nature, in conscience, through the prophets and apostles, in the acts of God to deliver Israel and in the life, death, and resurrection of Jesus Christ.[5] Revelation will be completed at the end of time when God's purposes are fulfilled but in the meantime the Holy Spirit will lead them into further truth. Revelation is also recognised in faiths other than Christianity. In Islam the Qur'an was sent down to Muhammad at intervals (surah 17.106) and he was directed to wait for it. In Hinduism there is a movement from the impersonal Brahman to a more personal Vishnu, and in Judaism belief developed in the one universal God (Isa. 45:5), who was everywhere both transcendent and immanent. When a revelation occurs or is said to have occurred it can be asked whether it is consistent with the belief system as a whole and free from contradictions. Does it make sense and cover human experience as we know it? Some revelations seem to fit the data better than others, so caution is needed.

But if nature reveals itself, as the scientist discovers, then we cannot try to approach it with our ideas and force it into our mould. Such a disclosure often reacts against our theories. There was great excitement when Newton said that light was like little bullets but the unexpected occurred when Huyghens asserted that it was more like waves. James Clerk Maxwell confirmed it but Einstein and Max Planck returned with the bullet theory! John Polkinghorne remarks: 'no-one wanted to believe that light was sometimes bullet-like, they were just driven to it by the way things are, and. ... even when we understand the physical world pretty well, looking at it more closely may show that it still has surprises in store for us'.[6] The debate was settled by Paul Dirac with his quantum field theory which explained that it depended on whether you were asking a wave-like question or a particle-like question. Like revelation in religion, nature surprises, shocks, and invokes a feeling of awe in the recipient.

Some see revelation as enhancement of human faculties by the inspiration of the Spirit. This may have a parallel with insights in science or a sense of rightness about something which is later confirmed by experiment. In religion questions arise about revelation: Is there one, where is it located, what form does it take and who has the authority to interpret it?

Emil Brunner insisted that natural theology is a bridge for special revelation, which is personal. God has revealed himself in history, in creation, and in conscience. Others stress the symbolic nature of revelation and say that it must be related to culture and artistic experience. Some like W. Pannenberg locate revelation in history and contend that it will only be fulfilled and verified in the end time. Most theologians believe that the Holy Spirit opens our eyes to the presence of God.

Those who argue that the scriptures are human projections about God and reflect what people wanted to believe and do are faced with the problem that God often calls on the recipient to do or say something which he does not want to do. The stories of the scriptures are full of this shock: Muhammad thought he was going mad, Isaiah was amazed (ch. 6), Jonah refused to obey and Arjuna in the Gita does not want to go to war with his relatives and reacts to the demand of Vishnu with awe and astonishment. It would appear that ideas of what God is like are falsified by intimate interaction with Him.

Revelation is progressive, for Jesus said: 'I have many things to say to you but you cannot bear them now' (John 16:12). In some passages in the Hebrew Bible it is thought that impure pagans should be eliminated but gradually it was seen that they should be converted, as in the story of Jonah, and in the New Testament. Some think revelation finished in CE 120 when the latest book in the Bible was written but that would rule out revelation in other religions such as Islam. Revelation is not limited to theological truth but appears in various areas of life, including poetry and inspired music: for instance, Handel's *Messiah*. He said when composing the Hallelujah Chorus that he saw the heavens opened and the great God sitting on the Throne.

But the interpretation of revelation is diverse and causes conflict. Some rely on the authority of Church or community or learned scholars. Within communities there is debate about it. In Sunni Islam there is the Qur'an and legal tradition through scholarly consensus but the Shi'ah rely on the authority of their Imams. In the mystical tradition of Sufism there is more latitude given to individual conscience and the idea of union with the divine. The Orthodox in Judaism rely on the teaching of the Pentateuch, the first five books of Moses, and are somewhat rigid in their interpretation especially the ultra-orthodox. But the Reform group relate more to modernity and accept the historical criticism of the Pentateuch. Other groups such as the Reconstructionists sit very lightly to revelation and have developed a kind of humanism.

But it is said that revelation is personal whereas science is impersonal. Michael Polanyi, as we noted, contests this point. He says that all knowledge is personal, involving commitment and involvement with that which is known. But it does seem that the 'I–Thou' idea of revelation is much more personal than the 'I–it' of science unless we are dealing with the human factor as in sociology.[7]

The authority of scripture

In general, writers accept natural theology, and Tom Torrance, though influenced by Barth, seeks to bring it into the sphere of revealed religion. His position shows a reaction to the historical criticism of scripture which tried to separate the message and the messenger by removing the theological frame of the Gospels because it was the perspective of the early Church.[8] There is a danger of interpreting anything by a thought process which is alien to what is being studied, in the desire for objectivity. This does not take into account the need for participation in the life of those we are seeking to get to know.

Torrance is raising an important point, for how the Bible is regarded will affect the view of revelation. For a long time now the New Testament has been subjected to rigorous scrutiny by form criticism. It developed from the 1920s onwards and focuses on the smaller literary units that have a function in the *sitz im leben* of the early church. A gospel might be of a form fitted for controversy with the Jews and in the form of stories generated for that purpose but not spoken by Jesus. There is also criticism which concentrates on how traditions have altered and grown. In addition redaction criticism considers the way the sources have been adapted into the final text.

We will see how this operates more fully in a later chapter but note here that much of this work has been praiseworthy, testing the evidence for the Faith. However, when taken to extremes, it tends to make the writers of the New Testament greater than the Founder himself. It is out of step with modern trends in the appreciation of literature which stresses scripture as story rather than history and it has sometimes jumped to conclusions which have later proved to be wrong. For example, it is now recognised that the Fourth Gospel may be historically more reliable than previously acknowledged. It used common Christian sources and is not a theological rewriting of the other gospels, as is seen when stories such as the feeding of the five thousand and the anointing at Bethany are considered. The Dead Sea Scrolls have shown that Jewish and Greek ideas were combined before the common era in a way that was once thought to be unique to John and of a late second-century date. Sites have been identified which John mentions, including Bethesda and Gabbatha, so stories associated with these may be historical.[9]

But in general, if science keeps the empirical and the theory together, as Torrance insists, then we must be cautious in separating them in the study of the New Testament. He points out that since Einstein geometrical patterns and physical structures in our spacetime universe are bound together so that conceptual and empirical factors inhere in one another. Abstract patterns are not imposed on nature but arise as nature reveals itself. Experiment and theory, empirical and theoretical factors interact with one another and must not be divorced, hence historical and conceptual ingredients need to be taken together. After the Enlightenment there was the observation of phenomena and by logical deduction natural laws were established, but this was reading laws into nature not from nature. Absolute space and time were imposed and the result was determinism, hence tribes in various countries were observed and interpreted by means of a mechanistic theory of evolution. Now we recognise the distinctiveness of the life and community of such people and allow them to convey to us their experience. We do not impose ours on them.[10]

The early church's interpretation of what Jesus was cannot be set aside on the grounds that it is not part of the empirical data, for in 1 John the empirical and theological elements are woven together. Jesus cannot be interpreted in terms of our own culture; he was a Jew and must be understood in the context of Israelite history. Another way that Torrance seeks to put his point is in his opposition to dualism. He laments the dualism between the sensible world and the intelligible world as in Plato or between appearance and reality as in Kant. If this is true then an unbroken relationship between Jesus and God does not hold but with the *homoousios* of Nicaea there is a oneness of being with God.

Singularities do occur in nature: black holes, for instance, so we must not write off singular events in history. Religions contend for the uniqueness of their founders and, in the case of Christianity, if God entered the world in Christ then both theology and science are dealing with spatio-temporal concepts. Torrance holds that theology has its own distinctive method, just like any other science. One distinction is that revelation judges us, not vice versa. It is *offenbarkeit*, an objective unveiling of God which removes our lack of perception due to sin. Torrance and Barth assume the actuality of revelation and then proceed to its possibility. James Barr says that Torrance is denying that God can be known from the natural order but it is more likely that he is contending for its subordinate role where revelation in Christ will judge it. The later Barth did come to recognise revelation or lights in the world though he qualified them by contending that they shone through Christ. Other religions who seem to be his 'lights' will find this difficult to accept and I have tried to find an approach which does more justice to them elsewhere.[11]

But there must be a test for revelation no matter how convinced the recipient is that he has received it. We can ask if it is logically consistent with what we already know about God or what problems it solves. The founders of religions insisted that those who sought to follow them must put their teaching to the test. Did such teaching enable them to master the body, and serve others? Did it help in time of suffering, enable them to conquer bitterness and apathy, to forgive, to co-operate with others? and so on.

Revelation is connected with religious experiences and these can be kept alive by the combining of stories and re-enacting in ritual. Past events become present, as in the Passover of the Jews and the Eucharist in Christianity. The present looks back to the past and forward to the future. But while many of the events in religions can be represented in rituals they cannot be repeated, which shows the difference with science. Yet science has its unique unrepeatable events, for example, the state of the universe in the seconds following the Big Bang and unique histories such as biological evolution.[12]

Community and problems

The community is important in both science and religion and individual experience is assessed by it. When discord occurs diplomatic measures try to resolve it but, if they fail, a split in the religion may occur. While there are differences between scientists, their community is more united than their religious equivalent and is more objective, rational, universal, and critical. It does not have the same kind of commitment as religion or use story and ritual, with religious models evoking attitudes and calling for personal transformation and involvement. While there is respect for history and recognition of past scientists, the idea of revelation in historical events contrasts with science. But religion does seek evidence for its beliefs. The followers of Jesus pointed to evidence in the Hebrew scriptures of a suffering Messiah and stressed that they had observed his ministry, death and resurrection and called for behaviour in accordance with his teaching.

Theology often emerges when presented, as with science, as a problem. How were the disciples and apostles to fit in their observation and experience of Jesus and the Holy Spirit with their belief in the unity of God? The eventual answer was the doctrine of the Trinity, which involved the risk of bold conjectures. Modern religion too is faced with many problems. Does it adapt, conform, or change radically its beliefs? The problems are concerned with both belief and practice. In the face of declining congregations, how can they attract worshippers without losing the hard core of belief? Is there an answer to the problem of suffering and evil? Can Christianity continue to sanction war? Is there a possibility of a fruitful dialogue with science? What can be said about feminism and homosexuality? Is there a model of God which will attract a scientific age? But the hard core must be preserved.

A theory, as we noted, starts with some problem and offers a hypothetical description of a wider situation, so that seen within this wider context the phenomenon is no longer a puzzle. We can, for example, construct a theory about the puzzle of suffering to show that belief in a good and powerful God is not an error since it is part of the larger situation of which our lives are a part. It seeks confirmation or falsification in human experience. Experimental testing occurs in the present but there is also hope that good will be brought out of evil. We consider this difficult problem in Chapter 6.

Religious experience is theory-laden. One of the reasons for the Jewish rejection of Jesus was that he did not fit into their ideas about the Messiah. How could the humble carpenter from Nazareth be their warlike leader? With Islam how could Muhammad's monotheism be acceptable in the context of the polytheism of Arabia? What was asserted did not fit into the beliefs and theories of those who heard them. Hence the cross and persecution of the followers of the prophet.

Science has both 'bottom-up' thinkers and 'top-down' thinkers and in recent times there has been a movement away from beginning theology with discussion of revelation to the problem of social situations. An example of this has been liberation theology both in South America and South Africa. This is the praxis approach arising out of the struggle for justice. It is the 'bottom-up' approach from the experience of the poor. Such a situation is correlated with what a religion teaches and whether or not it is really concerned about the plight of people under unjust regimes. Revelation must speak to particular situations and not be bound by dogmatic rules.

Some think of revelation merely as information but it is more than that according to the records: it is saving knowledge.[13] There is personal disclosure which results in transformation of the life of the recipient so it is both objective and subjective (Phil. 2:5). Often it is unexpected and concerned with real-life experiences. The prophet Hosea mourned because his wife left him for another man, yet he could not stop loving her, just as God loved unfaithful Israel. Saul on the road to Damascus discovered how surprising, sudden and dynamic revelation was and what transformation it could effect in his life. Personal knowledge is required so Jesus asked Pilate: 'Do you say this of me or did others say it to you about me?' (John 18:34).

There are paradoxes in scientific and religious thinking: the wave/particle duality and the divine and human in Christ. But religion deals with meaning, values, explanations, morality, whereas science is concerned with matter, energy, life, and so on. Science is concerned with proof but you cannot prove love, loyalty, justice, honour by the scientific method.

Testing

Experiment is the test in science and to a certain extent it is true of religion which tests faith. The assumption is made that God exists and it is contended that the assumption leads to an assurance for he who does the will of God shall know the doctrine (John 7:17; Heb. 11:6). In Buddhism, the Buddha insisted that those who wanted to follow him should stop asking questions and try the experiment. Faith is a common factor in life. How could we make any progress in personal relationships or politics or business without it? We trust the teacher to teach our children properly, the banker to keep our money, the judge to administer justice without partiality. Faith in religion is a form of insight, seeing the reality behind the appearance. It goes beyond reason and cannot be proved by reason. Some would say it was a sixth sense, a seeing of the invisible as shown by the story of Elisha who saw the unseen host surrounding the city while others only saw the armies of Syria. But once faith is exercised it seeks understanding. It is significant that Anselm, Aquinas, Descartes, constructed arguments for the existence of God from within the circle of faith.

Faith is always faith in something or somebody and rests on evidence which is considered reliable. It means believing and acting as Werner Heisenberg said:

> If I have faith it means that I have decided to do something and am willing to stake my life on it. When Columbus started on his first voyage into the West, he believed that the earth was round and small enough to be circumnavigated. He did not think that this was right in theory alone, but he staked his whole existence on it.[14]

Falsification as a test can be applied to religion, for traditional answers to the problems of today often do not satisfy. Creeds and confessions of a faith come under particular scrutiny and those who are ordained to the ministry of the Church often subscribe to them with mental reservations. Confessions can be treated as auxiliary hypotheses and modified or rejected. The method differs from the work of some theists who set out to confirm what they believe and set aside objections. The early church sought to falsify a philosophy like gnosticism which taught that Christ had only seemed to appear in the flesh.

In the next chapter we will continue to see similarities and differences between science and religion as we examine the language used.

3
God Talk

The problem with religious language is that we are dealing with a topic which is far above our concepts and understanding. Hence some philosophers have argued that God talk is nonsense, but others, on the basis of analogy, metaphor, symbols, models, myth, and story, assert that we can discuss the subject. The use and validity of such language will be considered and comparison made with the scientific. We will continue to argue that the methods of science and religion do overlap at times since the latter is concerned with the world and us. We have seen this in the last chapter when thinking not only of revealed theology but natural theology. There will also be a consideration of those religious thinkers who argue that religious language can refer to God and those who limit it to religious practice.

As we have said, the various faiths are based on scriptures and traditions and these writings make use of every literary device to communicate. The Bible, for example, is a library of 66 books which include most forms of literature: history, allegory, poetry, prophecy, laws, prayers, proverbs, symbols, myths, metaphors, and parables. There is a developing revelation and we cannot appreciate it if we apply only the test of scientific and historical accuracy. I cannot read poetry in that way or plumb the depth of a parable by skimming the surface or spending time on historical correctness. The scripture is poetical when it refers to the rivers clapping their hands and the mountains singing together and uses analogy in comparing the universe to a woman groaning in childbirth (Ps 98:7–9; Rom. 8:19–22). We can ask if the parable of the Prodigal Son is true or fictional but it does not matter because the meaning is the fatherly God welcoming the sinner with open arms.

Literal

Much trouble arises when we take some statements literally. The Jews believed that Jesus intended to destroy the temple but he of course meant the temple of his body. Nicodemus in his interview with Jesus thinks of physical rather than spiritual birth. The Fourth Gospel is full of symbols seeking to get at the deeper meaning of events and excels in demonstrating that earthly things represent the heavenly. In the Revelation of John we have the description of heaven which includes harps, crowns and gold. These are symbolic attempts to express the inexpressible. Crowns suggest that those who dwell there are sharing in the power of God while gold which does not rust represents the timelessness of the place and how precious it is. The harps indicate the ecstasy which we associate with music.[1] In the eucharist or Lord's supper, the loaf with its broken fragments is the symbol of his body and the cup the symbol of his blood, signifiying death but also the new covenant.

In the Qur'an, Allah is understood as very close to us, being nearer than our jugular vein. Symbols convey power and meaning with the material basis analysed by science and the conveying of ultimate value and power by religion. Thus Paul Tillich said that God is Being itself or Ultimate concern, something which gives meaning to everything. When the followers of the various religions observed their founders they saw something of ultimate value and the Hebrew prophets performed symbolic actions and did things to show the action of God. Amos speaks of a basket of summer fruit indicating the end of Israel (8:2) and an almond tree showing how watchful God is.[2]

The lamb is the symbol of sacrifice, a sin offering (Lev. 4:32) the paschal lamb for Passover (Ex. 12). It was the symbol of meekness, obedience and required protection (Isa. 40:11). Jesus is the lamb of God and even when the omnipotence of God is pictured in Revelation the lamb predominates over the Lion of Judah: God persuades and does not coerce (5:5–6). It is by the blood of the sacrificed Lamb that the martyrs conquer (Rev. 12:11).

Conflict with science was often due to taking the Bible literally as in the dispute over Ps. 93:1 where the opponents of Galileo held on the basis of a literal interpretation that the earth did not move. This owed more to the teaching of Aristotle than the scripture. Again, dispute arose about the passage in Genesis which said that God created everything 'after its own kind'. But it could mean that God acts in an orderly manner and not that 'kinds' meant 'species'. If this is so there is no need to oppose the theory of evolution on that ground.[3]

Metaphor

Metaphor means the transfer of a name from its original referent to another which is accompanied usually by a transference of feeling or attitude. Hence there is a transference from literal war to the metaphorical: Paul speaks of fighting the good fight of faith against cosmic powers which are the superhuman forces of evil in the heavens (Eph. 6:12–18). It is contended that all of the language used to refer to God is metaphor with the possible exception of the word 'holy'. It adds vividness to a discourse so we have the eye of the needle, mouth of the river, and so on. Metaphors are used regarding the death of Christ: justification, reconciliation, redemption, and expiation.

We use them constantly in describing something. I might say that the chairman of the meeting dogmatically settled the matter but it is prosaic compared to saying that the chairman cut through the discussion like a knife through butter! In that kind of description the metaphor gives us a model of the chairman who is authoritarian and brooks no disagreement. God is disclosed by metaphor for we cannot identify our words with his and biblical writers speak of him with awe and emotion. A metaphor leads to new insights, helping to understand or make clear the unknown and see the ordinary world in a novel way. It is a way of knowing, not simply an insight.

There are personal metaphors of God: father, mother, husband, friend; but also impersonal ones: rock, water, fire. The Bible draws them from the human and non-human. God is a sun with a voice like a torrent or thunder and his justice is compared to the deep oceans. There is the animal imagery: God is like a lion, panther, leopard, bear, he carries us on eagle's wings and protects the nestlings (Ps. 84:11, John 3:8).[4] On the human level God has head, face, eyes, ears, mouth, but he does not have a body of flesh (Isa 31:3) and anthropomorphic imagery is used very sparingly. Moses cannot see God's face (Ex. 33:23), Isaiah sees the Lord enthroned (ch. 6), Ezekiel perceives a human form but it is only the likeness of God (Ez. 1:26–8). Yet he is very much in action, seeing, speaking, punishing, healing, guiding, and shows love, pity, patience, generosity and justice.

The metaphorical can be true or false and its referent can be real or unreal. Some think for example that the church as the body of Christ is a true metaphor in the ontological sense while others deny it. If comparison is explicit we call it a simile and meant to be taken literally, if it is implicit it is metaphor and non-literal but this does not exhaust the difference between the two. When two things are compared they are not alike in all respects as when we say that people are like sheep and need a shepherd.[5]

There is no reason for saying that the literal came first and then the metaphorical for there is a mixture from the beginning in the biblical writings. Caird finds fault with R. Bultmann in postulating the writers of the books of the Bible as prescientific, and thinking that modern man would be more comfortable with existentialism. There is picture language in the Bible but it is an assumption to say that the writers understood it as literal fact. Attempts to reconstruct the mentality of primitive man have very little to do with the understanding of biblical literature.[6]

Metaphors shape meaning and are cognitive. They give rise to models like 'father', which are extended metaphors.

Models in turn lead to theories and can create a new paradigm. Christianity has a basic metaphor, the kingdom of God which has arrived with the coming of Christ but will be fully realised in the future. The problem was that the followers of Jesus understood what he was saying about the kingdom literally and thought of a materialistic one. Today, some Christian theologians compare God with a field of force but in physics that is a material thing.

Models, analogy, symbol

The Hebrew model of God was a mobile one, moving with the Israelities in their journey through the desert and going with them into exile. He cannot be confined to a temple and even the heaven of heavens does not contain him.

Science uses models: computer, scale, thought, and practical models. There is the geometrical model of the universe possessing four dimensions, three of space and one of time. Electrons are imagined to go round a central nucleus like planets round the sun but this model proved inadequate for hydrogen atoms which are not solar systems. Every model has positive features of likeness but negative aspects of unlikeness.[7]

Paradoxes appear in both science and religion. The latter teaches that we are good but bad, made of the dust of the ground but in the image of God, individual creatures yet social, possessing a body but also a soul. Christ is divine yet human, God is transcendent yet immanent, and so on. In science light is both a wave and a particle. Science uses analogy, metaphor and similes. It speaks of the flow of electricity and light as waves but these are not literal descriptions. Science employs analogy in connection with theoretical models and these are often based on creative imagination. The understanding of a new concept may be helped by comparing it with one that is well known and the new model may suggest an application to new phenomena. Models are like illustrations shedding light on what is new but they must be tested. When a paradox arises, complementary models may be called for, as with the quantum wave and particle.

Such models are limited ways of imaging what is not observable. The critical realists, Ian Barbour and Ernan McMullin, hold that models give us insight into real structures and are justified by their success.[8] But the world of quantum is strange and it is difficult to visualise what is going on.

Darwin proceeded by analogy when he thought of nature selecting the fittest. He studied pigeon-breeders and how the breeder selects and rejects so as to develop desirable characteristics and he hit on the idea that this was analogous to what nature did. Natural selection was a metaphor and illuminated a continuous process. But selection implies purpose, choice, and intelligence and the analogy seemed to imply that Darwin was personifying nature. He insisted that what he meant was the aggregate action and product of many natural laws issuing in a sequence of events which we could discern. But in appealing to something known, pigeon breeding, he was transferring to nature something not intended. What he and subsequent biologists wanted to avoid was any evidence for design.

Science makes use of models which go beyond observation and no theory is in agreement with all the facts. It is naive realism to think that we have in the laws and theories an exact replica of nature. The models, analogies, symbols, are unable to point to all the features of an object but they represent realities in the world and are not just useful fictions. They are neither simply based on sense observation nor identical with mental states but approximate to the truth of reality. Evidence can be produced, for example, to show that atoms and electrons are real, but as in religion, the models are limited and partial and do not describe reality as it is in itself. We know how the world behaves, not what it actually is. Both in theology and science, models are revisable and subject to change.

Concepts in science are expressed in language or mathematical formulae which point to structures in objects which are hidden from empirical investigation. Analogy relies on a similarity but difference between entities. Thus the wave theory of light developed on the basis of an analogy with the wave properties of sound: some of the characteristics were similar but others different. In the same way when we speak of God on the basis of what we are and how we act we recognise similarity but difference. With regard to science a model could be called a systematic analogy because we are moving from some law which is understood to another which we hope to know.

Mathematical models are symbolic representations of physical systems used in order to predict behaviour, but theoretical models spring from creative imagination. The concepts of science are symbols and very indirectly refer to the atomic world because there is no direct observation. Of course every discipline is interested in some particular aspect of an object and only describes it from that viewpoint. A musician listening to a symphony marvels at the texture of harmony, form, themes, but a scientist hears it as a set of molecular vibrations. Analogies are used from familiar situations to aid understanding before models are expressed in mathematical formulae. They are subjected to rigorous tests and some are discarded.

There are many models of God. He is designer in Genesis 1 bringing order out of chaos, potter (Jer. 18:6), architect (Job 38:4), builder, farmer, shepherd, hero, warrior, doctor, judge, king, husband, mother and father. Such language is used carefully and there are warnings about it, for God is not like man (1 Sam. 15:29). But the spiritual cannot be identified with the abstract, hence anthropormorphisms are indispensable. God is Lord and King of the world and manifests his Wisdom and Word through creation. In the prologue to the Fourth Gospel the active rational creative principle is incarnated in Christ. The model of potter and a craftsman assumes a static product and is not suitable for a world and humanity which are constantly changing. Important models are father and mother, showing discipline and compassion, and God as Spirit means vitality, creativity, inspiration, illumination and guidance. God has also been understood as an artist or poet or dramatist or agent suggesting unpredictability, intention and interaction with his medium.

If the Spirit works throughout the world and in all areas there may be some parallel with what Kuhn says about intuition. He thinks of it as a 'lightning flash' enabling an obscure puzzle to be seen in a new way. But the puzzle does not occur in a vacuum, for it relates to the science of the past. Its context is that of the old paradigm but it is not linked to certain items of it. There is a transformation. In religion the model of God is seen in a new way but building on the past context. Kuhn uses terminology which is familiar in religious discourse to describe these experiences, that is, 'conversion'. But resistance to new paradigms persists because of a lifelong dedication to an old one and the fact that arguments for the new one are not decisive: Einstein's resistance to the work of Heisenberg, for instance. Thus the Jewish religious leaders in the first century clung to their picture of God, resisting the insights of Jesus, and the Arab tribes initially opposed Muhammad's view of Allah.

Model of divine scientist

We know that the universe and its laws can be understood and captured by mathematics. In the sixteenth century Johann Kepler advanced the understanding of the orbits of the planets and pointed out the success of mathematics in capturing the patterns of the world and the correspondence of the human mind to that of God's:

> In that geometry is part of the divine mind from the origins of time, even from before the origins of time (for what is there in God that is not also from God?), it has provided God with the patterns for the creation of the world, and has been transferred to humanity with the image of God.[9]

And Galileo Galilei believed that mathematics was grounded in the being of God and that the universe cannot be understood without it. Today Roger Penrose thinks that mathematical laws are not human constructs but written into the universe which implies the mind of a scientific mathematician behind them. Penrose writes:

> There is something absolute and God given about mathematical truth' and 'not only is the universe "out there" but

> mathematical truth has its own mysterious independence and timelessness'. Paul Davies agrees: 'The laws of nature are real objective truths about the universe and we discover them rather than invent them.'[10]

Keith Ward understands God as the cosmic creator or intelligence and sees no reason why such an intelligence should not make itself known in a personal way. It must have a purpose, as shown by the experiences which people have had down the centuries. He is the active power that inspires the universe to realise new actualities from many possibilities which are inherent from its origin.[11]

The glory of God is the stamp of divine artistry which the creator has impressed on all his handiwork (Ps 19:1). In his answer to Job he refers to the foundation of the earth and how he measured it and laid its cornerstone. The picture is of the mathematical architect and builder but the one who continues to care for his creatures, shown by the reference to the mythical monster behemoth (Job 40:15). We can look for rationality in God and in his dealings with us because the scripture says: 'Come let us reason together says the Lord ...' (Isa. 1:18) and Jesus told his hearers to love the Lord their God with all their mind (Mk 12:33). It is true that God's thoughts are higher than ours but not totally dissimilar.

The model of God which we wish to develop is the divine experimenter who has given freedom to the subjects of the experiment and potential to develop. God is experimenting and improvising in an open-ended way through continuing creation. There is a framework of law in evolution but also the element of chance. The scientist uses experiment to test his theories and it would appear that this world is a testing laboratory employed by the divine scientist.

The model will not replace the traditional ones but interact with them as the authority of kingship interacts with that of a compassionate father. It rules out any thought of an impersonal scientist, for the personal is necessary in many ways but particularly in worship. We would not address God in that context: 'Dear cosmic scientist'! Models need to be qualified or taken together with others to try to get a picture of God. Thus 'father', as the feminists point out, needs to take into account the mother model. Models do reflect the culture of the day so that 'shepherd' was very appropriate for a rural society and still conveys to us care, protection, guidance; but now we need in addition something that reflects the primacy of science.

The model for the Christian is Christ and he has a cosmic dimension relating not only to humanity but to all things. According to the apostle Paul, Christ is pre-existent, lord of creation, sustainer of the world, and reconciler of all things (Col. 1:15–20), the image of the invisible God and the fullness of the divine essence (Col. 2:9), and ultimate goal of all creation. Paul's view ties in with the model of the Logos by whom all things came into being (John 1:1–4, 14). Logos means rationality or order and relates to the pattern and structure of the universe. The Word was in the mind of God so the writer of the Fourth Gospel follows the thinking of the Greeks but also the Jews for the Hebrew word *dabar* means action. God not only spoke the word of creation which was in his mind (Ps. 33:6) but acted or became incarnate in Jesus Christ. If Christ has the mind of God then we can discover what God is like through him.

The Logos in the Fourth Gospel parallels wisdom in Hellenistic Jewish thought where both were associated (Wisdom of Solmon 9:1–2). The creative presence uniquely present in Christ became more than verbal, a human life, an explicit statement of incarnation. Logos theology developed in the second century and became more rationalistic than in the Johannine writings and figured in the development of the Trinity. But because it appeared to make Christ a second God it was replaced by 'person', that is, hypostasis.[12]

We might add at this point that there is a need to get away from the caricature of the scientist who has an indifference to human life. In applying the model to God we perceive his intention to realise a personal relationship with the subjects of the experiment. The believer speaks of God, faith, grace, because she knows the effect on her life. Her faith is tested by her behaviour. Images of God are similar to scientific models, as symbolic representations of what is not observable, but they differ from the scientific in calling for personal commitment, values, and have a direct relationship to worship and behaviour. There is an ultimacy which is not reflected in science. Models of God, while they reflect more than human ideals, must at least measure up to them. Various societies embrace different values but at least five are fundamental: freedom, justice, knowledge, wisdom and happiness. A religion which opposes these is suspect.

The history of science shows the replacement of one model of the world by another, and ideas of God develop in religion, so a high moral deity is announced by the Hebrew prophets, and in Hinduism there is a movement from the impersonal Brahman to the personal Vishnu. Tests for models of God would fulfil the scientific criteria of rationality, internal coherence, fitting with the data, comprehensiveness, and fruitfulness. An image of God as king, for example, has given the impression of coercion and needs to be brought into relation with mother, lover, shepherd, husband, and so on.

Most theologians agree that we need a model of God which does justice both to the majesty of God and his personal nature. Extremes produce deism, a God who is remote and uninterested, or One so human that he looks like a projection of our desires and wishes. We think that the model of the cosmic scientist can bring together the impersonal and personal in God, for the view of him as the Mind behind the universe receives its full personal dimension in Jesus Christ. Later we will see that we might view him not only as the suffering servant of Yahweh but as the suffering scientist.

But this means embracing the incarnate model of God and it is significant that there are incarnational strains in some religions. Even when a religion starts with little reference to the gods, as in Buddhism, there is a development of personal devotion so that veneration is given to the Buddha and the Bodhisattva in the Mahayana form. And it may be said that heavenly Buddhas are incarnate in the human Buddhas, that is Gautama Buddha, with the next expected one being Maitreya, the Kind One. There are similarities between the Buddha and Christ, yet differences. Both taught that we are self-centred and need enlightenment, neither tried to explain the world that they saw as transient and temporary, and both lived lives detached from those things which tend to dominate ours. But Jesus did not advocate withdrawal from the world and he was conscious of his need for God. The image of Christ which has predominated is of a suffering saviour but the Buddha after his enlightenment lived serenely, had many followers, was successful, and died peacefully at the age of eighty years. Christ died cursed, despised, a false prophet and blasphemer. The image of the Buddha reflects serenity compared with the crucified Christ, yet suffering is the major factor in Buddhism. Mahayana Buddhism introduced the element of compassion and proceeded to develop the incarnational element in the doctrine of the three bodies of the Buddha.

Despite the Qur'an's opposition to the sonship of Christ, the Shi'ah also developed the incarnational element, as did Sufism. It is also present in certain elements in Sikhism. Divinity can be used in different senses. Muhammad and Nanak are charismatic leaders and have an intense relationship with the divine. With the Buddha there is no relationship with God but he is the spiritual principle of Buddhahood, a channel for Dharma: the Teaching. We might think of the Buddha as the 'window' into Dharma, so that the human body, the physical individual, Gautama, virtually ceases to have any importance. The doctrine in Christianity is not easy to explain and there have been heresies and deviations from the orthodox belief. In Christianity the divine becomes man, but in Hinduism it is not clear that it occurs, for avatar means 'coming down': the divine manifesting itself rather than becoming human. The models in Hinduism see Vishnu as father, friend, lover; and in Islam he is Protector, Forgiver, Bestower, Provider, Forbearing, all Forgiving, Generous, Loving, Giver of life, Source of all Goodness, Pardoner, Compassionate and Guide. In all the religions with exception of Buddhism there is the dual model of personal and impersonal.

Religious models attract more personal involvement, as noted above, than do scientific models, though a scientist can find a model so dominating his thinking that it is difficult to give it up. Models being closely related to analogies are usually based on something known but they can soon be discarded in science if they do not fit empirically. Feminists see limitations today in the male model of God, and the kingly one is not liked because of its authoritarian connotation. But perhaps we need to modify and reinterpret them rather than abandon them.

Myth

As well as models, religion makes use of myth and legend with the former used in a technical way in order to convey meaning. History and myth are intertwined but the disadvantage of 'myth' is that it often means an untrue story. Perhaps it would be better to use 'story' and recognise that there were many stories concerning creation in what is called primordial time. A myth is not a judgement on what is true or false but tries to express the inner meaning of an event or person. Jesus, for example, was an historical figure but Christians understand him to be more than that, namely the Son of God.

Scholars disagree about the definition of myth. Rudolf Bultmann defined mythology as 'the use of imagery to express the other world in terms of this world and the divine in terms of human life, the other side in terms of this side'.[13] But does the Jewish belief in a three-decker universe, as he asserted, mean that Jews believed that God lived in the sky or does the upward look mean reverence for his majesty, transcendence, glory and so on? They knew and state plainly that heaven cannot contain God (1 Kings 8:27) so they appear to be viewing heaven and earth symbolically.

John Hick defines myth as a story which is told but which is not literally true or an idea or image which is applied to something or someone but which does not literally apply but invites a particular attitude in its hearers.[14] There can be narrative myths which fall into this category and mythical images or concepts used to identify and show the significance of a person or situation. It is this significance which is more controversial than using the term 'myth', for Hick thinks that incarnation stemmed from a later interpretation of the early church. Their context was that of divinity applied to the Roman emperors, so an easy transition to Jesus was likely. But did not the early Christians reject worship of the emperor and often lose their lives as a result?

Hick admits that there is no agreed way of using the term and there are true and false myths. We say 'He is a devil', which is not literally true but a vivid description of bad behaviour. Similarly the creation myth is not literal truth but mythologically true; it is saying that the world is a divine creation and we are imperfect and live in an imperfect world. Sorting out what are true and false myths leads Hick to deny the existence of Abraham which would not endear him to those religions, which as he says, express the Real (another problem), but look back to Abraham as their father.[15]

We agree, however, that 'Son of God' in connection with Jesus cannot be taken in any physical sense but is used metaphorically. However, even if we set aside the Fourth Gospel with its claims for the uniqueness of Christ, which Hick wants to do, there are in the other gospels 'thunderbolts from the Johannine heaven' and Jesus did what no other prophet had done: he forgave sins. On the positive side Hick recognises Jesus as one of the most God-filled beings we know but somewhat ambiguously denies the orthodox view and seems to think the myth can be accepted as God loving the world and sending his son to redeem it. But such redemption is limited to our overcoming self-centredness. He says that Jesus would have regarded as blasphemy the idea that he was God incarnate.[16] But the charge against him *was* that of blasphemy, if we are to accept the synoptic records.

I think that the term myth needs to be used with caution and the scripture is not happy about it because of its fictional nature (1 Tim. 1:4, 4:7; II Tim 4:4; Tit. 1:14; 1 Pet. 1:16). An extreme has been reached in some theology where myth is used to cover the whole range of theological language. Of course there is value in myth when used properly. Some see myths in science: Newton and the apple, other universes, visitants from outer space, and the string theory. Myths can be connected with religious rituals and social movements. Plato understood them as conveying universal truths and connected with origins, and Jung believed that they expressed permanent needs in our view of the world.

Other literary devices

In addition to myth, religion uses literary devices that are not found in science: parable, allegory, irony, legends and so on. Most scholars insist that 'in the parables one stands on the bedrock of authentic Jesus tradition'[17] and we will have cause to refer to them in later chapters. Paul introduces allegory into Christian exegesis (Gal. 4:21–6) and Philo's interpretation of the Torah was entitled, 'Allegories of the Law'. It was also employed by Origen and the Alexandrian school. Irony and humour, sarcasm and satire figure in scripture with the picture of God laughing at the pride and deceit of sinners. Irony is saying one thing and intending another. The Second Book of Samuel records David dancing and Michael commenting: 'What a splendid display the king of Israel made today' (6:20). Jesus describes comical people (Lk. 1:5–8), comical ideas (Mt. 5:36) and comical events (Mt. 5:15).[18]

There can also be historical legends based on fact but embroidered in stories of the patriarchs in Genesis or aetiological stories which explain the name of a place or customs associated with them. Beer-sheba, Bethel, Israel, and so on. Saga too is used, being the weaving of legends into connected wholes, as in the Homeric poems or the Icelandic sagas. Hence scholars have seen in the various writers (J, D, E and P) of the Pentateuch divergent versions of a common epic narrative or saga. The story of creation, the accounts of the antediluvian patriarchs, the Flood, and the Tower of Babel, were brought from northwestern Mesopotamia to the West by the Hebrews before the middle of the second millennium. Alan Richardson in his commentary on Gen. 1–11 refers to them as parables rather than myths.[19]

The scriptures also employ hyperbole or overstatement. Thus we read that Jacob loved Rachel more than Leah ... the Lord saw that Leah was hated (Gen. 29:30–1). It was a preference, and reappears with God preferring Jacob rather than

Esau, and Jesus saying that if anyone comes to him and does not hate his father and mother he is not worthy of him. The use of extremes was common to the Semitic people and attracted attention: taking the plank out of your eye, gulping down a camel, and so on.

The use of religious language

We now consider briefly whether or not religious language refers to God or is to be limited to its use and practice. It is the debate between realism and non-realism in a religious context. I have mentioned it in some detail in a previous book and refer here only to the main aspects.[20] D.Z. Phillips believes in the non-realist view that God exists in religious communities. He is influenced by Wittgenstein who argued: 'Don't look for the meaning, look for the use.' Language is like a bag of carpenter's tools, each with its own particular function or with a range of games each with its own rules and criteria. Thus we have the language of science, legality, economics and so on. Language is social and cannot be uprooted from the life of the community in which it originated.

A stranger in a foreign country needs not only to master the rules of its language but to study its culture, customs and traditions, if he wants to understand its people. Only when we see the use of language can we understand the meaning. The language of religion is intelligible within a mosque, temple, church, or gurdwara, and is subject to internal rules. It does not make religion a private affair, for the community has rules to regulate the behaviour of its members, and they will govern their conduct in the market place. Wittgenstein did not think that God was unverifiable or non-existent but inexpressible. But this means that religious statements have no authority except within the faith community.

Swinburne opposes Phillips and sets up the usual arguments for the existence of God but Phillips says that often these arguments result in an impersonal God rather than, as in the ministry of Jesus, care and love for the poor. God cannot be defined, for he is inexpressible and if we try to capture him in doctrinal statements he becomes one existent among others and contingent like us. God cannot be inferred from the world but known by his role in the worship of a community and in that context. He refers to Wittgenstein's 'picture' theory of language, and points out that the religious believer's image of the world is different from that of the unbeliever. They disagree about how 'to picture' the world rather than on matters of fact. It is a difference of preference and commitment due to different cultures and means that we can never grasp the truth or get at the real meaning of anything. In religious practice the reality of God is assumed without any attempt to demonstrate his existence by reason.

The view that God is the explanation of the universe is set aside since it would mean that he was in competition with other explanations. Phillips comments ironically on those who believe in the probability of God's existence and have to change the creed so as to recite: 'I believe it is highly probable that there is an almighty God, maker of heaven and earth'!

Don Cupitt agrees with Phillips and says that Christianity has portrayed God as a powerful Ego, self-existent and omnipotent, the Great I Am. It is incompatible with the Buddhist stress on the illusory nature of the ego and with what the religious traditions teach about our becoming less egotistic and more selfless. Cupitt will have little to do with metaphysics so he denies an objective God and immortality. We can go on worshipping in our various communities but our worship should be directed towards the ideals and values which we hold; God is the personification of our values. Since he believes that language is unable to grasp certainties we deal with appearances, not realities.[21]

There has been much debate about Wittgenstein's 'language games': about language used by various communities, and whether or not he meant that it separated them and that they could not judge one another. With him the role of philosophy is descriptive but it also must be critical. Such criticism shows that all frames of references or views of the world and God are not equally good. Karl Popper opposed Wittgenstein, arguing that we not only criticise communities but break out of them. This is the realistic view. Of course we need to know how language is being used and the coherence of the whole system which would be one test of its truth. But non-realists like Phillips and Cupitt say that language is relative to the culture in which it is used and we must not look for correspondence to reality. It fits the beliefs of a society at a particular time but later it is shown to be incorrect. At one time people believed that the earth was flat in accordance with their world-view but it is not within our frame of reference. In such a changing world we never have reality.

It is true that religious traditions change in their expression of the faith in different historical and social contexts but a test is how a contemporary communication is continuous with the beliefs of its founder. Muhammad, Jesus, Nanak, and the prophets of Israel, all believed in the existence of God in some objective way and would hardly have recognised equating him with our values. The fact that revelation comes through the human does not mean that it is simply the expression of the human, for it can oppose natural instincts and beliefs.

Phillips argues that God is real for religious communities but not outside them. Yet many people are religious and acknowledge the reality of God but do not belong to any particular community. A more serious problem is that such a proposal separates science and religion since the first is seen as objective and factual, predicting and controlling nature, whereas the second is subjective, relating to beliefs and aims at a particular form of life. But we cannot talk about the purpose of God in the world if his use and function relate only to particular religious communities. The scientist holds that there is an external world apart from his practice of trying to understand it and we would contend that God exists apart from religious worship.

It is true that one form of life cannot judge another by its own terms but truth is universal. Philosophy will critically examine all statements whether religious or not and there is no special sphere in which it is not allowed to work. Cupitt and Phillips want to make religious statements expressive or exhortative and not referring to some external reality. Cupitt, stressing the inability of language to grasp God, argues that no certainties about him are available, but if he believes such a statement, labelled perspectivism, to be true then he recognises a certainty! On the other hand if it is merely a point of view why should we take it seriously? If religious experience is relegated to the subjective and has no objective referent we are agreeing with Freud and Feuerbach.

There is a dispute between Phillips and Cupitt regarding language as to how it might refer but both appear to make God into a personification of our ideals. John Hick disagrees with Phillips' view on this point because it means that God does not exist as a factual truth. Phillips uses religious language which presupposes the existence of God but he fails to see that philosophy is engaged in clarifying whether such a claim is justified. It would appear now that he, in the light of criticism, is prepared to give up thinking of religious language as an autonomous language-game cut off from other beliefs, but he seems to have no room for a life after death being in the purpose of God.

Having considered some similarities and differences in the mode of expression of science and religion we examine in the next chapter the beginning of the divine experiment.

4
The Experiment Begins

Both science and religion are interested in the beginning of the universe and the development of life but, as we noticed, it is said that science deals with the how of creation whereas religion tries to answer the question why. In this chapter we begin with the how and then proceed to the why.

The origin of the universe

There are 100,000 million stars and galaxies and it is reckoned that the universe is about 12,000 million years old. We arrived a few million years ago which was somewhat late on the scene. The planets were formed by a swirling cloud of dust and gas which gradually fused together. Stephen Hawking argues that the initial beginning of the universe was influenced by quantum effects; it was self-contained and did not have a beginning or end. Hawking is not an atheist but cannot believe in a God who is interested in us in the way religions proclaim. He notes that belief in God requires faith and thinks if there was a God he could act in ways that physical laws cannot explain.

Some of his scientific colleagues oppose his view that the universe is self-contained and believe that God both creates and sustains the universe. They argue that talk about a beginning has no relevance to its creation. An artist's line may have a beginning or end or form a circle but that has no relevance to the question of its being drawn. Hawking seems to leave the matter of the existence of God 'open ended' for he realises that his mathematical approach to truth is only one way and there are other ways to truth which are valid. He thinks that time melts away when we get back to the Big Bang but if time did not begin with it how can we have a quantum fluctuation from the initial state of nothing? Such fluctations that we know of occur in space and time.

Earlier, Hawking and Roger Penrose had put the stress on a singularity at the beginning. It is a mathematical point of infinite density and is at the centre of a black hole. Hawking decided to apply the singularity to the whole of the universe. Stars collapse inwards and form a black hole and it would bend spacetime round upon itself just like a running track. The black hole would then be cut off from the rest of the universe and nothing could escape from it. There is a surface to the black hole like the surface of the sea which would indicate the boundary between the universe and it. The universe could be a black hole with gravity holding everything together and spacetime forming a self-contained closed entity which folded round on itself. But black holes pull matter inwards towards the singularity whereas the universe is expanding outwards from the Big Bang.[1]

Space expands and we think of it as a rubber band which stretches with the spots on it moving apart. This pictures the movement of the galaxies. We think of moving through space but that is incorrect, for the spots do not move through the rubber band! It was the Big Bang which stretched space. Everything in the universe was squashed together and then there was a sudden explosion with all the bits flying apart. The flash of light from it is still here but cannot be seen, just as we cannot see the waves in the microwave oven.

The anthropic principle

Much use is made by theologians of the anthropic principle. It appears that we were meant to be here for there is the regularity and uniformity of the physical environment which enables life to evolve. There are four basic forces: gravity, electromagnetic, weak and strong nuclear. These forces are close in their values and the ratios correct otherwise we would not be here. The masses of the particles and the ratios of the masses required fine tuning to produce the right conditions. The earth is the only region which moves in the solar ecosphere and the only region which is neither too hot nor too cold. The other planets (Venus, Mars, Mercury, etc.) have extreme conditions, so it is unlikely that life ever existed on them. But currently Mars is being thoroughly investigated in the hope of finding water. Jupiter 'the king of the planets' is lethal with its volcanic eruptions and deadly cold.[2]

It is not only the immensity of the universe which is astonishing but how all this fine tuning was done. The density of matter had to be just right after the Big Bang and gravity tuned to slow down the expansion. Hydrogen and helium had to be fused to get the elements of carbon, nitrogen and oxygen which are the building blocks of life. The heat temperature in the interior of stars made this possible and these stars were formed by gravity working on clouds of gas but it needed to be right otherwise the stars would not have ignited. Only when the level of carbon and oxygen had been finely balanced could life evolve. Physicists point to what they call a resonance level which is an energy state inside the carbon nucleus that aids the process. Had the level been 4 per cent lower there would not have been enough carbon for organic life on earth and had the similar resonance level in oxygen been only half a per cent higher most of the carbon would have been turned into oxygen. Either way we would not be here.

Theists do not say that the anthropic principle is a clear argument for the existence of a superintelligence or designer but a pointer in that direction. If such an indication is correct it would be possible to assume that human beings are the goal of God's creation. Those who oppose the anthropic principle contend that there could be many universes and we are lucky enough to find ourselves in this one. It was just an accident. Ironically, the cosmologists who believe this do it on the basis of faith for there is no way of proving that such universes exist and even if they did they are forever invisible. In any case, if belief in a designer God is speculative, so are the other solutions of the result of chance, random fluctations and so on. We cannot however make an easy agreement between the Christian doctrine of creation, and the Big Bang scenario, for science is not certain about an absolute beginning.

Hubble pointed out that the universe was expanding in a dynamic mode. We know that in order to launch rockets we need a speed faster than 11 kilometres per second to escape the gravity of the earth. The same applies to a universe with a critical launch speed at the start of its expansion. If too fast gravity will not halt and it will expand forever but if too slow there is a contraction to the initial state. It is amazing that our universe had the right critical launch speed, for without it there would not have been stars and the possibility of our existence. And the universe expands in a well-ordered manner proceeding at the same rate in every direction. What is the cause of this? John Barrow speculates that there might have been some external influence present or a principle of symmetry or economy.[3]

The remarkable fact is that the hot density at the start was in a state of equilibrium and determined precisely the different particles in the universe. Before the universe was a second old the numbers of protons and neutrons were equal. Taking this into account it is hard to avoid the thought that there is a rationality behind the universe. Theists will see a connection here between what physicists are thinking and the logos of the Fourth Gospel. The theory of inflation, however, explains the expansion and uniformity of the universe not by some principle which demands order at the start but by antigravitating states of matter which act as a brake. But what kind of matter? There is the existence of dark matter which is invisible but could have acted in this way. It exercises a gravitational pull on stars and galaxies and may be composed of neutrinos carrying no electric charge, and not be subject to the electromagnetic or nuclear force.

Chaos and superstring

However, uncertainty remains about the beginning of the universe, for at that point we are beyond the conditions which we can simulate on earth and uncertain about the laws of nature involved at these energy levels. We do not have a complete theory of the elementary particles of matter and the forces that govern them. The first seconds of the universe were dominated by quantum uncertainty and so far gravity and the other three forces of nature have not beeen united in one theory. But there are chaos theory and superstring theory, with the former contending that systems which obey precise laws can nevertheless act in a random manner. What is revealed is a strange universe where such things as circles and ellipses give way to complex structures known as 'fractals'. Even a small disturbance like a butterfly's wing can change a chaos system like the weather! It is not a clocklike world, for clouds are not clocks. Chaos is widespread and it raises questions about predictability, measurement, and the verification or falsification of theories. We have irregularities in nature but it may be that they are obeying laws which we have yet to discover.[4]

The string theory promises a unified descripton of all the fundamental forces of nature and will, it is hoped, result in a theory of everything. The idea is that the physical world is made out of little strings instead of particles and arose from the investigation of scientists into hadrons which are the interacting particles that emerge from high energy collisions in particle accelerators. Hadrons contain quarks and their force produces bonds which are like pieces of elastic joining the quarks. The interactions of quarks and the hadron resemble a whirling string. In the string theory there are eleven dimensions and this has resulted in many ideas. Critics said there was confusion, but in answer it was argued that the theories were different ways of looking at the same thing. The strings of energy vibrate like a great symphony playing the tune of the universe. There are parallel universes but we cannot see them and the theory is not testable. It contends that the Big Bang has occurred before and there will be many in the future. The theory has struggled to gain recognition over the years but some scientists still regard it as mythical.

A theory of everything would predict the key properties of mass, electric charge, magnetic moment, describe all the interactions between particles, and give an account of the fundamental forces of nature and their strengths. It would tell us more about the dimensions of spacetime and how the universe came into existence. The search is motivated by the belief that nature ought to be simple.[5]

Understanding time is also a problem. Is it an unchanging background in which events occur, so creation is in time? Or is it connected with events and created with the universe? The latter appears more acceptable and theologians are keen to point out that the idea was put forward by Augustine. But what is the nature of time: is it absolute (Newton) or relative (Einstein)? Theists are happy with creation out of nothing and time starting with the universe but scientific theories in connection with creation are speculative since we cannot carry out experiments. The only hope is to try to correlate them.[6]

But some scientists argue that matter and energy arise from nothing, a random quantum fluctuation producing an equal quantity of matter and anti-matter. But where did the quantum laws come from which made it possible? Perhaps the quantum fluctation is the way God creates out of nothing which in this case has properties. It appears that scientific explanation is within the framework of reality whereas theology is talking about the ground of reality itself.

Design

Theists believe that an accident or chance or coincidence could not be an explanation. How can such purposelessness produce humanity which has all kinds of purposes? Is the atheist saying something meaningful when he says that the world is meaningless? Sir James Jeans in *The Mysterious Universe* wrote: 'We discover that the universe shows evidence of a designing or controlling power that has something in common with our own individual minds. ... We are not so much intruders in the universe as we first thought.'[7] A scientist delights in the laws which describe the regularity, the kind of laws which would be the work of a Mind producing a world which is intelligible. For, as it has been said, if it takes intelligence to work it out, might it not have taken an Intelligence to have put it in place to begin with?

Before Darwin it was David Hume who was thought to have delivered a decisive blow against a universe which had been designed. The argument from design is an analogy because it insists that the universe resembles designed things within it and must have a cause like theirs. Hume asserted that any order could be the result of chance and the world might be an organism rather than a machine. Further, there are signs of disorder as well as order. Why should there be one designer rather than many? Arguing from design leads to an infinite regression. If the universe is unique we would have no experience of such a cause or causes as we have with human designers.

But to think of an organism rather than a machine does not rule out order, as we see in ourselves. With regard to chance, theists accept it but also necessity, which is a framework of law. Chance is regarded as the means of exploring and experimenting and it can have a cause. It may be thought that when I get a cold it was chance but I will seek around for a cause. Hume initially said that we can imagine a beginning of existence without any cause, but we can comment that there is a difference between imagination and reality. Eventually he seemed to be saying that he had been misunderstood and asserted that he had not meant that anything might arise without a cause. Finally, a material object cannot be the cause since it would be part of the universe itself.[8]

The founders of natural science were men of faith: Copernicus, Kepler, Galileo and Newton, who believed in a rational creator. He had given the laws or rules of the universe which can be stated in mathematical terms. This implies a mathematical scientist behind them. Paul Davies insists that the universe is no accident but structured in a way that provides a meaningful place for us. After surveying the alternatives he concludes that the impression of design is overwhelming and that while science may explain all the processes whereby the universe evolves its own destiny it still leaves room for a meaning behind existence. Davies thinks that mysticism goes beyond logic and science but he postulates a 'natural God' within the universe not a supernatural interventionist one. Yet he recognises the mystery of the universe and that answers lie beyond empirical science.[9]

Keith Ward throughout his book, *Chance and Necessity* (1996), insists that there is a cosmic mind, transcendent as well as immanent, whose goal is to produce values of goodness, beauty, truth, and the triumph of virtue, beneficence, compassion and love. The victory will not go to what evolution stresses, the exterminators and replicators, but to those who cooperate in the realising of values of different kinds and care for their environment. The cosmic mind desires a personal relationship with us and a sharing in his knowledge and creativity and will give us the grace and power to achieve it.

But Hawking's belief that the universe is due to quantum fluctuations means a randomness which would oppose design. Quantum theory, which raised questions about scientific determinism, presents a problem for order. His work indicates that the universe has no boundary and is self-contained and not affected by anything outside it. It would just exist. God would not be needed to create or to tune the laws for its evolutionary purpose. But theists reply that God is not simply an event at the beginning but is actively engaged in the world and that the no-boundary proposal has not gained universal assent. It does not explain why the universe exists. Hawking seems to be arguing against a first cause which is temporal and can be disposed of, since quantum theory has questioned the causal explanation by showing that the behaviour of individual electrons and other small particles is unpredictable.

It is interesting to compare two statements. Dr Peter Atkins, who takes an anti-religious position, writes: 'The universe can emerge out of nothing, without intervention, by chance'; but Darwin spoke of the 'impossibility of conceiving that this grand wondrous universe, with our conscious selves, arose through chance'.[10] As M. Poole points out, we often say that something happened by chance but if we knew all about it we could conclude that it was planned. Scientists speak of chance and randomness about events which cannot be predicted, such as individual particle random activity but the behaviour of large numbers are predictable. For example, large numbers of randomly colliding gas molecules fit in with a regular pattern of behaviour which is described by Boyle's Law.[11] Chance and law seem to work together, as we will see in a later chapter with regard to evolution.

Why is there a universe?

The question calls for a theological answer but is it consonant with what science is saying about the How?

It is asserted that creation means that everything depends upon God and is not about temporal beginnings, hence it is not possible to infer directly from the Big Bang to creation: 'if our universe had a beginning in time through the unique act of a creator, from our point of view it would look something like what the Big Bang cosmologists are talking about. What one cannot say is that the doctrine of creation "supports" the Big Bang model, or that the Big Bang "supports" the Christian doctrine of creation.'[12]

God is not some kind of physical cause to be placed on a par with the Big Bang but is the explanation of why there is something rather than nothing. According to Genesis and other passages of scripture (Job 38:1–41; Ps. 74:13–15), creation is the work of God. As we mentioned these are not literal accounts but parables and stories in the context of the ancient Near Eastern mythology which depicted God's victory over chaos but Genesis transforms such myths, for the forces of chaos or formless reality are not divine. They are different from the myths in the fertility cults. But it could be thought that God's defeat of the forces of chaos is a prelude to creation and it continues with leviathan whose chaotic powers have got to be kept under control (Job 9:8, 26:8–13; Ps. 89:8–11).[13]

The model of God is that of a master builder (Ps. 127:1) or an artist. But if creation is from nothing (the idea of creation from nothing first appears in 2 Macc. 7:28) what does it mean that he created mankind from the dust of the ground and where did the formless chaos (Gen. 1:1) come from in the first place? Chaos is the dragon which is subdued but not finally eliminated until the end and a new creation comes into being (Isa. 27:1). It is a continuing threat represented by the serpent in the Garden of Eden who urges the humans to claim autonomy, knowledge and power. The episode in the garden is a test and we are reminded that testing is a key factor in science. They fail it, are ejected, and experience spiritual death which means separation from God. Another way of looking at chaos or the *tohu wabohu* is to see it as empty turbulence out of which God created heaven and earth (Gen. 1:2; Isa. 34:11). And it might have some kind of parallel with the quantum fluctuation in physics.

There are two accounts, P (from the fourth century BCE) and J (ninth century BCE) in which God creates beings that he can communicate with because they are in his image. J (as we can tell from its use of Yahweh as God) is a folk tale and older than P (priestly writer). There are other writers involved in the composition of the Pentateuch but the literary theory has been questioned since it was introduced by Julius Wellhausen in the nineteenth century. However that may be, J has the insight that humanity is part of the natural order. God breathes into man and he becomes a living being not an immortal soul hence a psychosomatic unity. Since the writing of the accounts took place after the exodus from Egypt it is clear that Israel knew God as redeemer or liberator before thinking of him as creator.

The word for create is used only of God (Isa. 4:2–6; 42:5) for it is something which the human cannot do. God hovers like a mother bird over the newborn world (Dt. 32:11) having called it into existence. Creation is good and man is the crown of creation. In P the animals are created after their kind not individually, which throws light on the species question. Man is mortal like the beasts but made in the image of God which means that he can obey or disobey his creator. It contrasts with the stars that move mechanically and the animals who follow their own instincts. Man chooses wrongly and the image is defaced but not obliterated and is transmitted to posterity (Gen. 5:1, 3; 9:6) Male and female created together are complementary in P not as in J with the female coming after the male.

There have been many interpretations of Genesis but commentators are agreed that we are dealing with symbols: the special tree, a talking serpent, a forbidden fruit. Humanity comes from the dust of the ground in J and life begins in the sea in P. The idea connects with the evolutionary theory of origin from the pools in which the first replicating molecules originated and in P there is the ascending scale from the lower to the higher which is mankind. Creation is continuous, as if the divine scientist was exploring possibilities. The conditions are right for life but the divine gardener does not make the plants grow. Humans are called to develop the world in all its aspects.

The means of creation, developed in the apocalyptic literature, focuses on the concept of Wisdom, which existed before creation (Prov. 8:22–31). But Genesis stresses the Word of God, which is seen as dynamic in Ps. 33:6, 107:20; Jer. 23:29; Isa. 48, with God active in laying the foundations of the earth. The Targums (translations of the original Hebrew) tried to remove anthropomorphic references to God and stressed the role of the Word. The Word produced the regularity of the universe and rationality in the mind of mankind. Philo said that the Logos was the thought of God stamped upon the universe and in the mind of man. From logos we get logo and logic, hence the Word of John 1:1 is linked with a logical explanation of the world created by it. The writer of the Fourth Gospel wrote in the context of Greek ideas and linked the Word or rational principle of the universe with their view of the Logos. He was communicating with the world of his day and it is what we are trying to do in speaking of the cosmic scientist as a model for our age.

Philo's point is significant, for it is amazing that we are able to understand the mathematical laws which govern the operation of the universe. Einstein commented that the only incomprehensible thing about the universe is that it is comprehensible. We can stand back and see what the rules are even though we have come from the dust of the ground or in the words of the physicist, animated stardust. How can this be other than the thought of God stamped not only on the universe but on our minds?

But the writer of the Fourth Gospel did what the Greeks, with their belief in the evil of matter, would never do. He spoke of the Word becoming flesh in Jesus Christ. Plato held that in the unseen world there is the perfect pattern of everything, the doctrine of forms, with the Good as the highest, being the pattern of all patterns. On earth we have the copies of these forms but Christianity saw Christ as the archetypal Word with the whole cosmos patterned on him. He is not a copy but the real pattern come to earth.

Stress was laid on his humanity as well as his divinity (John 4:6; 8:58). However, a distinction is made between God and the Word, for God is *theos* and the definite article is used, *ho theos*. There is no definite article with the Word so the noun becomes more like an adjective. The Word is not *ho theos*, that is, identical with God, but since *theos* is used it means the Word was of the same character and quality and being as God. Hence in Jesus we see what God is like, though in incarnate mode. The basic claim of the writer of the Fourth Gospel is that the mind of Jesus reflects the mind of God, the words of Jesus are the words of God and the actions of Jesus are the actions of God. When challenged Jesus pointed to his works as witness to who he was and showed a rationality that defeated his opponents. He justified his action of healing on the Sabbath by pointing out that they circumcised on the eighth day, which could fall on the Sabbath (Lev. 12:3). Moses their lawgiver had laid it down that there must be no work done on the Sabbath day and yet they attended to medical needs which were not necessary to save life. On another occasion, when accused of blasphemy, Jesus referred to gods mentioned in Psalms 82:6. He pointed out that if God could address as gods those to whom the word of God came, that is, those who held positions of delegated responsibility, he cannot be guilty of blasphemy in claiming to be God's son when he holds a commission from God.[14]

The explanation of the world cannot be limited to the causal since that does not tell us why it is intelligible whereas to postulate God means purpose and significance. But despite the rationality, purpose and design, humans turned away from God, according to the Genesis record, and he repented that he had created them. The parable of the flood shows the frustration of God because of the evil impulse or evil imagination (*yetzer ha-ra*) inherent in man which leads to idolatry and unchastity.

The divine scientist is grieved at the failure of the experiment and reacts emotionally. But it is not a total failure for through Noah and his family the experiment is recreated and in subsequent history of Israel there was always a remnant who remained faithful to Yahweh. The renewal is confirmed by a covenant (*berith*) made with Noah and later with Abraham and eventually with Israel at Sinai. God's overall purpose will not be thwarted by the disobedience of mankind.

Kenosis

According to Christianity, however, the greatest renewal was done in Christ and we can understand that the Word in becoming flesh limited himself in various ways. The same would apply to God since he has created mankind and given us freedom, but it is a voluntary limitation. The passage discussed by recent writers is Philippians, Chapter 2, which traditionally is referred to as the kenosis or 'emptying' of the Word.[15] They begin by pointing out that exegetes are divided over whether the scriptural passage refers to the pre-existence of Christ and his giving up of that state or refers to his self-sacrifice. But if the first view, which is traditional, is accepted it can be extended to God, so that in creating he qualifies his power and knowledge and becomes involved in suffering. This would result in being subject to change and affected by the world.[16]

Ian Barbour, who is influenced by process theology, contends that there are human choices which are unknowable until they are made, so God does not know them. He says that God is Lord, majestic and awe-inspiring (Isa. 6) but also like a sorrowful husband forgiving the unfaithful wife (Hosea 1–4). Israel is his suffering servant bringing redemption to other nations (Isa. 53). Usually it is thought that the limiting of God's power was a voluntary one, but Barbour contends that it was a metaphysical necessity. By this he means that the limitation of divine power is inherent in the divine nature and not a kenotic act. This is debatable.[17]

God is considered to be vulnerable (Vanstone) and some express it in the trinitarian interaction, but Barbour follows process theology. God acts through his Spirit and the dove-like image indicates non-coercion. But we might ask, is it not true that the Spirit is also symbolised in wind and fire, which signify power? In process theology God is fellow sufferer and there is the dual nature of God: the primordial source of all possibilities, and the consequent, which is influenced by the world. Charles Hartshorne, another process theologian, defends dipolar theism: God is temporal and changing in interaction with the world but in himself eternal in character and purpose. This God of persuasion, in the light of the past, orders potentialities and novelty. His nature has a hierarchy of levels with downward causation and information from higher to lower. There is human freedom though genes do constrain choice.[18]

A.R. Peacocke thinks that the freedom of the creature places a limitation on the creator who suffers. He believes in panentheism, a feminine model, and thinks it is confirmed by Paul in Rom. 8:19–22 where the whole creation is in labour pains. Creation will be set free from its bondage to decay, and will obtain the freedom of the children of God. The creator is vulnerable because free persons can obey or disobey, hence God took a risk in creating. Jesus Christ is the supreme example of this risk and vulnerability. Peacocke argues that God is immanent in the process and has endowed nature to unfold itself. Such a world must involve pain and death as the old gives way to the new.[19]

Concerning the passage in Philippians, Peacocke says that Paul is quoting a hymn about the crucified man 25 years after the death of Christ. He is relying here on C.F.D. Moule, *The Origin of Christology* (Cambridge University Press, 1977). The later thinking of the church was based on the uniqueness of Christ already present from the start, so the development was a drawing out and articulating of what was always there. Peacocke stresses his openness to God and his pre-existence which for Paul was personal. But he thinks it refers to the purpose and intention of God. Jesus is the result of one particular initiative of God, a new emergent, a possibility always inherent in man, hence possible for all of us. But having said that, he admits that we may not be able to dispense with the ontological significance of Jesus. The DNA code of Jesus was patterned on the same genetic code as ours but the openness to God and his relationship with him led to his sacrifice for us. Consequently, in his resurrection he has been taken into a new mode of life within the very being of God.[20]

Jurgen Moltmann insists that the kenosis means a rejection of the unchangeable, impassible, immovable God of philosophical theism. Creation is a limitation for God as he allows space for his creatures and he chooses to suffer with them. We know that love makes room for another. Holmes Rolston agrees with Peacocke and points to the text: 'unless a grain of wheat falls into the earth and dies, it remains alone but if it dies it bears much fruit' (John 12:24). He contends that selfish genes must be tempered and corrected by self-actualising organisms. Redemptive suffering is a model that enables us to understand nature and history but true kenosis is not found in nature because there is no voluntariness. Only the human can choose the truly altruistic preservation of the other at the expense of the self. But altruism in nature is still debated.[21]

John Polkinghorne now believes that God allows divine special providence to act as a cause among causes. In the life and death of Jesus we see an interweaving of divine and human causes. God does something new and unique in the resurrection of Christ since it was only expected at the end of history. Other writers see kenosis as a unifying theme for life and cosmology and God's action as creator. Kenosis is joyous giving up of selfish desires and making sacrifices. It is clear in the case of Christ.[22]

It is clear that what these thinkers are saying in various ways affects their view of the attributes of God. His immutability means that he is steadfast, faithful in character and purpose, but includes movement and change, for he is dynamic. God does suffer but he is not controlled by passions which destroy. Suffering is not a sign of weakness but strength. God is omnipotent but in the sense that his sovereign love has no equal. He is love which means vulnerability and risk, for love can be rejected. Omniscience is not simple knowing all but deep wisdom which includes the foolishness of the cross. Philosophical speculation without attention to what God has done in Christ is not helpful.

But other scholars have contended that God cannot be fully or in every respect represented in human form. What does 'fully' mean? If it means that his divine glory and majesty were veiled, we agree, but would believe that he was revealed in accord with the limitation of the form. There is a subordination of the Son to the Father in the New Testament but it cannot mean inferiority. We do not say that a son is inferior to his father even though he shows his subordination by obedience. An analogy is of the piano soloist playing a concerto under the direction of the conductor. She and the other players are subordinate to that direction but they are not inferior; it is merely that their roles are different.

But apart from the extension of kenosis to God the scriptures do not hesitate to speak of the limitations that he has imposed upon himself. In the course of events it becomes apparent when he is surprised at the reaction of Israel (Jer. 3:19, 5:7), and amazed that she would worship Baal (Jer. 32:35). The people do what he does not want (Isa. 54:15) and he at times changes his mind about places like Nineveh that he intends to destroy (Jonah 3:10). This human model shows how like man God is, yet unlike him in his patience and love. The essence of love is to give and when we give ourselves completely to another person or strive to realise some goal we feel 'drained', 'exhausted' and 'emptied', but we do not cease to be what we are. Similarly God remains what he is and retains his omnipotence, using it when circumstances demand. Thus the power of God is shown in the resurrection of Christ and in the control of nature (Job 38–41) and will be exercised in the consummation of all things.

The act of creation itself involved omnipotence. What do we mean by this attribute? It is generally agreed that God does not do that which is contrary to his nature or the freedom which he has given his creation. He wants them to love him and respond in a voluntary way which can lead to suffering if the wrong choices are made. God cannot do what is logically impossible for that would involve him in contradiction. But he could use his power at some times such as creation and not at others, that is, intervening in the execution of Christ. We cannot dictate to God when he intervenes or interacts with his creation. There is the power of love and it is the reason for the cross in that by it Jesus intended to draw all men unto himself.

In taking this attitude to his omnipotence we are thinking of an everlasting suffering God, not a simple, timeless one. With the latter view of his omniscience it is like someone on the mountain seeing all events simultaneously so he knows all about our future. But if this were so, are we not determined? It would make the knowledge of God causal and oppose human freedom. Perhaps it is better to say that God does not know what our particular choices will be though he can predict as Jesus did the betrayal of Judas. God has been likened to a Grand Chess Master who has overall knowledge of the game but does not know each particular choice. We cannot thwart his plans and he will eventually win the game. He remains in control for he is God and we have all the limitations of the human. There is an interaction between God and us so that he continually acts to bring us back to obey his plans. This is a personal God involved in time which opposes the anti-realist view of a simple timeless God.[23]

In conclusion, theism contends that the world is dependent upon God for he chose the initial conditions in order to make life possible. God could have made things to make themselves and that would mean that he has endowed matter with potentialities to develop more and more complex organisms. But the doctrine of creation differs from the scientific account which relies on the physical sciences explaining how the world works, whereas the theological account tries to understand its meaning or purpose. Science cannot prove or disprove a divine purpose in creation but the same applies to any action. An action can be understood in a physical sense as due to the movement of muscles and so on but as to why the action took place needs the examination of the inner motives, intention and purpose of the agent and these are difficult to determine. Perhaps the motive of the agent was greed or doing good or deception or religious. How and why then results in seeing science and theology as complementary.

In the next chapter we continue the theme of creation as we consider the human subjects of the experiment.

5
The Subjects of the Experiment

In this chapter we need to understand what evolution teaches about us, the subjects of the experiment, and how far the dialogue between religion and science can help in understanding the divine experimenter. It is necessary to mention some of the basic points made by Darwin and then go on to see the development of neo-Darwinism. It will then be contrasted with what the scriptures teach about creation and the nature of fallen humanity. Darwin's theory opposed the designer argument for the existence of God for it meant descent by modification, with natural selection as the new creative force. It motivated adaptations and changes in species and was a mechanical process operating just like the stars in their courses, hence there was no teleology or purpose. But we would like to stress again that in this experiment there is no case of the subjects being treated as objects as in many scientific investigations. We are subjects who are personal and we will constantly refer to the divine experimenter's love, compassion, and longing for our response.

Evolution

All organisms go back to a single cell, and natural selection works on small inherited differences to produce in time forms that are different from the ancestor species. There is competition between species for food and mates and the fittest survive because they have inherited good characteristics. It leads to gradual changes in species but Darwin was not clear on how this occurs though he realised that the geographical isolation of a species was important.[1] He placed stress on observation but he is likely to have had the theory in his mind, for Erasmus Darwin had put forward evolutionary ideas. Such ideas were circulating since the publication of Robert Chambers' *Vestiges of the Natural History of Creation*, in 1844. But thinking about evolution goes back to the Greeks, Empedocles and Aristotle, and there was also the Christian belief that God could have imparted potential to matter and left it to develop: Gregory of Nyssa (331–96). Augustine too thought that such potential would lead to life with the details working themselves out in accordance with the laws laid down by God. But from the sixteenth century onwards it was special creation which dominated.

Darwin's case rested on variation among species, the struggle to survive and reproduce, and that useful properties occur in variations which are inherited. Nature will select those with the best characteristics. In his early thinking he saw these as the means which the creator used but he gradually lapsed into agnosticism about any action of God in the process. The theory has received much criticism over the years: the fossil record is discontinuous, gaps are common, some forms remain the same while at times new forms appear quite suddenly. The record is imperfect so that transitional forms only appear for an instant, as is seen in the transition from invertebrates to fish. Only the dominant species are recorded. Some biologists believe the process is gradual; others contend for punctuated equilibrium: jumps or bursts. Darwin said that the process had to be gradual otherwise his theory would fail, and neo-Darwinism insists upon it. But if punctuated equilibrium is favoured it gives the theists the hope that God might be involved. What we do know is that there have been many changes in the forms.

The origin of life remains a mystery since spontaneous creation is not an option, as Pasteur proved. He discovered that microbes were already present in matter, not generated spontaneously by it. Life comes from life, but evolution depends on reproducing organisms. Where did they come from if no creator? Could life appear by chance? Despite the confidence of some biology textbooks on how it originated Francis Crick says that it is 'almost a miracle' and reaches out to space for an answer.[2] In any case even if God employed natural forces to originate life, there is no reason why his involvement should be ruled out.

The religious response takes various forms, including rejection of the theory, but this is difficult to maintain in the light of the scientific evidence. A better approach is to see evolution as the method of the creator who creates the conditions for life as we noted in the last chapter and gives the world freedom to develop itself. Religion answers the questions: why and who created, science the what, when, and the how. At one stage in his thinking Darwin said that the creator originally breathed life into a few forms or into one and it led to endless forms being evolved. But his belief suffered a devastating shock when his favourite daughter died, and he ended life as an agnostic. A.R. Wallace, who arrived at the same conclusions as Darwin, found it difficult to accept that our unique characteristics, intelligence, moral and religious dimensions, could have naturally evolved, and argued for intervention by God in the process: 'a superior intelligence has guided the development of man in a definite direction, and for a special purpose, just as man guides the development of many animal and vegetable forms'.[3]

The DNA (deoxyribonucleic acid) discovered in the last century is an example of how biology has advanced since the time of Darwin and Wallace. The DNA is a genetic program inherited from the past and the coded information generates the characteristics of the organism and the means for its reproduction. It was made from copies taken from our parents' DNA so we look like them, being built from the same set of plans, but it is a mixture of the parents and not an exact copy of either of them. If mistakes were made in the copies or bodily chemicals affected or there was radiation we could have a brand new code, making us very different. The code is selected naturally. Neo-Darwinism then, taking its cue from Darwin, argues that evolution is a gradual process of small genetic changes brought about by mutations and the recombination of genes acted on by natural selection. Genetic studies have shown the plasticity of species and that the genetic code is essentially identical in all organisms. Life can be traced back to a single origin.

Genes

The discussion of genes has become commonplace. A gene is a unit of heredity in a chromosome, controlling a particular inherited characteristic of an individual. The chromosome is a structure which occurs in pairs in the cell-nucleus and carries the genes. Chromosomes are strands of DNA. In 1953, Watson and Crick demonstrated the double-helix structure of DNA, the genetic code which transmits the information of life. It is a beautiful twisted ladder and is made of building blocks of different types: A, G, C and T. Together they are in long strings and the DNA stores information depending on their order. Just as a vast number of words can be made from the letters of the alphabet so DNA can generate quantities of genetic instructions from its simple code.[4]

Genes convey DNA information to proteins that create chemicals such as dopamine, which makes us feel happy and excited. Too much makes us into extroverts, too little into introverts. Different proteins control the three hundred brain chemicals which determine how we feel, think and act. It is the DNA which tells the cells what to do and this includes the brain, which is designed by genes from both parents. Genes are wrapped up in the DNA. A gene is a stretch of DNA and some contend that our genes make us what we are but if this is so it makes them purposive. Sociobiology stresses the genes but in its weaker forms a place is given to the impact of culture. Cultural innovation affects the survival of the genes and alters their strength. It contrasts with the stronger version of sociobiology which teaches that mind is created by genes and culture stems from it. The weaker theory is more acceptable since it gives a greater independence to nature and nurture, and culture can act in opposition to genes.[5]

Richard Dawkins says that genes survive one body and occupy another, so their survival and the DNA are what are important. It happens by slow stages hence Dawkins' opposition to punctuated equilibrium. He embraces weak sociobiology for there are memes, which are cultural items. Examples of memes are tunes, ideas, fashions and so on, which replicate. The fitness of genes lies in their ability to reproduce but Dawkins does not answer what gives survival value to memes. Dawkins sees religion as a virus that must be stopped but others hold that as long as religion protects the genes by its ethics it will survive. John Bowker contends that the mistake which is often made is to think of genes and culture as causal forces.[6]

There is more than genes/culture affecting an individual, namely the total history of her development. A variety of causes are required for complex human beings, and personal choice involving freedom cannot be neglected. We are not only the result of genes and culture because we take control and select not only for survival but values which are good or bad. These may enhance gene-replication or not. Hence there are two kinds of fitness, the kind that we acquire in order to survive and also that which enhances life. There is the physical world of regularity which includes genes but also the world of values. The two are not identical though they do overlap.

Today we realise that we have fewer genes than we first thought, about 30,000, and we are learning that it is the combinations of genes, not a single gene, that shapes us. It is too simple to think that we can cut out a single gene and put another one in and all will be well. Genes are a secular version of original sin causing imperfection but we can transcend our genes. They are selfish but there is altruism: ants sacrifice to save colonies, the fireman risks death as he goes into the burning building to save the child and the soldier dies for his country. Natural selection aims at reproduction of genes but there is also sex for pleasure and contraceptives are used to prevent children. Evolution describes how we orginated physically with cruelty and waste but we think that it ought not to be and frame ethical codes of behaviour. Religion lays stress on ethical codes and has helped societies to survive. We have developed a culture which encourages the portrayal of beauty, love of music and art, and moral ideals. There is much more to life than survival, and sacrifice shows that moral ideals are real.

The case for neo-Darwinism rests on replication of the DNA sequence, random mutation of the sequence because of copying 'errors' or mutations in the sex cells, the struggle for existence with the better adapted surviving, and natural selection of the adaptive mutations.[7] Mutations, which can be good or bad, occur and are a change in a gene. It means that the instructions are altered, but mutants are rare. They enabled life to survive and produce varieties from a single species and the varieties in due course become species. The argument is that without such mutations *homo sapiens* would not have evolved, but it does seem difficult to accept that we are here partly because of mistakes in the DNA coding! Small mutations occur when pieces of the genetic code get scrambled as the genes are being copied either by one letter replacing another or by a chunk of DNA being accidentally cut out and spliced in somewhere else or a piece of chromosome getting inverted. Just as a misprint in a book can change the meaning of a sentence, so a copying mistake in DNA changes the nature of the organism that carries it. Mutations convey advantages: for example, ability to run faster and have a better chance of survival and reproduce. Thus we have variations within species and natural selection works on them, preserving the best. Scientists are now asking: why wait for random mutations to produce new life forms?

Genetic engineering, which has caused much debate, may be the answer. Genes have to be organised so that they resemble computer files of information but the materials used are not disks but the DNA. There are ordinary genes and super-genes, with the latter exercising control. Twenty-three chromosomes contain the genes but there are two centre chromosomes: XX for a female and XY for a male. Some females have a Y chromosome yet are born female because Y does not function. The problem can be corrected by an operation and hormone treatment. Genes can now be transferred from one animal to another and there is little difference between our genes and that of the chimpanzee.

The transmission of culture from generation to generation has an effect on the evolution of a people and it moves much faster than genes. Hence learned values affect our behaviour. Evolution may be seen as a change of gene frequency in a population of organisms. It was thought at one time that genes were completely separated from the cells of the body and could not be influenced from outside. But Howard Temin in 1971 discovered that viruses can transport genetic material into host cells and embed it in the host DNA so that it will replicate itself. The viruses manufacture a special enzyme to do this. Temin received the Nobel prize in 1976 for the discovery. It is accepted then that outside agencies can affect genetic changes.[8]

Purpose

Evolution is random but finely tuned and raises the question of whether there are just little blind leaps in the dark or whether it is being used as a device for creative possibilities. Is it an accident that the dinosaurs, who if they had survived would have wiped out self-conscious organisms, were killed by a meteor? Why is the process structured in such a way to a good end, namely *homo sapiens*? Is it just a coincidence? Does the existence of God render the process more probable than it would otherwise be? Evolution proceeds from a state where there are no values to one where there are; does this not reveal purpose? There is not only suffering in the world but beauty, cooperation and altruism. It is difficult to argue that consciousness, morality, rationality, science, religion and art can all be explained by the survival of the fittest or reproduction of genes, and Darwin did admit that natural selection is the main but not the exclusive explanation.[9]

Biologists in general avoid questions about purpose but Alister Hardy says that modern biologists have stressed the mechanical role of external forces acting on random mutations and neglected internal drives which can modify evolution. Animals are curious, display initiative, self-adaptation, instinct and learning; hence we can speak of the 'psychic life' of the animal, which is a most powerful creative element in evolution. Ian Barbour contends that it means purposive behaviour as well as chance mutation.[10] Yet some scientists contend that organisms consist of nothing but atoms and neglect the higher levels where novel forms emerge which are not predictable from the lower. With these forms we have purposive behaviour and mental life, self-consciousness, and a nervous system which unifies experience.

Natural selection is not a chance agency for it preserves improbable combinations through generations but it is blind, with no long-term goal, according to Richard Dawkins. Selection is based on survival or reproductive success. We are survival machines blindly programmed to preserve our genes which are selfish molecules. It would mean that we are biased towards our own interests which agrees with what religions say about sin, but they also assert that there is a goodness in us. Other biologists do not agree with Dawkins, arguing that whole organisms and species have a priority over genes.

Dawkins also contends that our brains predispose us to believe in a Designer since we design things ourselves. Hence we oppose the idea that complex design could arise in an evolutionary way from primeval simplicity. But having disposed of this brain illusion he said that we must accept the brain's teaching when it tells us that evolution ought not to be the way it is: our moral sense. The question then arises: Why accept what the brain says on the latter and treat the former as an illusion? Theists generally reply to Dawkins by stressing both chance and necessity, novelty and regularity, which God has given to the world. But there is, as Vardy points out, the drive towards complexity inherent in the evolutionary process. How can this be, if survival alone is the sole criterion for success? On that basis single-cell organisms could survive effectively, yet evolution is driven towards greater and greater complexity, resulting in the emergence of intelligence of a high order. There seems to be no reason why this should be so and the idea of an intelligence driving evolution might be plausible.[11] Intelligence explains what physical laws cannot: why nature plans in advance for our needs and we not only adapt to our environment but transform it.

It might be thought that if the genes are selfish they would be fighting one another for their place in the organism and so prevent complex parts of our bodies such as eyes, hearts, or consciousness, from existing. It is what Dawkins calls the paradox of organism and he seeks to resolve it with the genes combating their selfishness and combining to produce the necessary parts. But this means cooperation for the good of the whole organism. On this basis one would think that he would accept that the organism is the unit of selection but he continues to try to save his solution. What is significant is that he admits that there is something special about the wonders of the eye which would never have been achieved by 'almost all random scramblings of the parts'.[12] And what is said about the eye could equally well be said of basic components of our internal cellular and developmental machinery: the genetic code, protein assembly, nerves, consciousness, sexuality, language, and so on.

Dawkins is continually asking for evidence, yet he writes: 'The theory of evolution by cumulative natural selection is the only theory we know of which in principle is capable of explaining the existence of organised complexity. Even if the evidence did not favour it, it would still be the best theory available.'[13] Yet he refuses to believe in the activity of God because he says there is no evidence for it. Commenting on this, Gabriel Dover says: 'Scientific arrogance should not be used as a weapon against the supernatural. So used, it simply demeans the honesty and acceptance of the transitory nature of knowledge, which is at the heart of science. We know that physics has moved on from Newton and that biology has moved on from Darwin, but we are all still scratching at the best produced by evolution.'[14]

In contrast to Dawkins' view, the genes can be compared to a written description of building a house or ship. They are written in a code of a few symbols but the order can carry an infinite number of meanings. There is every evidence of planning making it difficult to accept that it was all due to chance. How could a disorganised collection of molecules assemble themselves into the whole of a living organism? Biologists admit it is still a mystery. An example is the embryo in the womb, which develops from a single fertilised cell into a living being. In the process many cells have worked to form liver and bone nerves at an astonishing level of accuracy: Even if damage occurs, new cells can replace mutilated ones; those that have got out of position can find their way back to the right place. Hence, 'in studying the development of the embryo it is hard to resist the impression that there exists somewhere a blueprint, or plan of assembly, carrying the instructions needed to achieve the finished form. In some as yet poorly understood way the growth of the organism is tightly constrained to conform to this plan. There is thus a strong element of teleology involved. It seems as if the growing organism is being directed towards its final state by some sort of global supervising agency.'[15]

Dawkins believes that he has found an answer not only to the how question but also the why. This is the ultimate question, the purpose of life. The purpose is not only to pass on our genes but to free ourselves from the selfish gene and enjoy art, amusement, and avoid reproduction by the use of contraceptives. We could break Darwin's rules as soon as our superior brains developed and do what the other animals cannot. Thus the adaptation of the environment, the use of language, cooperation, foresight, technology, led to our mastery of the world. This is non-genetic evolution and led to asking the question: Why are we here? The answer lies in purposeful humanity, not in God. This is is humanism but derived not from the lack of purpose in nature but mankind. The brain is sufficient explanation of our purposive behaviour but he neglects to tell us what mind is or what it does. He is saying that purpose is seen only at the higher levels of humanity, which can be disputed, but now he admits that culture plays a role. His memes are rather weak in that connection since they are little bits of information which are struggling to be replicated.[16] A better view is that cultural practices affect evolution and this is being stressed by his critics.

Darwin himself believed in 'designed laws' with the details left to the working out of what might be called 'chance' and it may be that such laws are the laws of physics and chemistry. Gradualism was a law of evolution and so was natural selection. Today evolution proceeds by natural selection, molecular drive and neutral drift. Molecular drive is a process capable of changing the average genetic composition of a population through the generations, and genetic drift is an alteration in its genetic constitution.[17]

Chance for the theist is the way God is exploring nature. It works in a framework of necessity that is the fundamental constants: laws of matter, energy, space and time. Chance actualises or explores the potentialities but its operation at the level of mutations does not prevent law-like behaviour with populations or organisms. God the cosmic scientist uses chance as his 'search radar sweeping through all the possible targets available to its probing'.[18] Arthur Peacocke argues for continuous creation with new kinds of reality emerging in evolution which reflect a living, dynamic God. He is not simply sustaining the world in a static way, for he has established an inbuilt creativity. New forms emerge with the fall of the dice or chance depending on the rules of the game. All of which means that God takes a risk.

Karl Popper would agree, since he saw propensities for certain properties to appear. Peacocke contends that natural selection favours them and at the level of *homo sapiens* there appear complexity, organisation, information-processing and storage ability. Such ability is the condition for consciousness but it means an increased awareness of pain and suffering, which are warning signals. These spur us to action and have survival value, but there is also the benefit of social cooperation. He thinks that, taking these into account and the impact of the environment, we can speak of a divine purpose without seeing it as deterministic. But there is no need for a special action of God such as intervention in quantum or chaos.

Prigogine and Eigen have shown the interplay of chance and law leading to the emergence of living structures. Peacocke prefers the metaphor of God as a composer unfolding the potentialities of the universe and creating a spiritual harmony, and refers to the Hindu god Shiva, the Lord of the Dance of creation. He uses the idea of the play of God in creation which I do not think is a good one since it does not reflect the suffering involved. But he can rely on the image of Wisdom playing at the creation (Prov. 8:27–31).

Since evolution implies the survival of the fittest it is difficult to explain altruism. Is morality simply rules for survival built into us by our genes? Where did we get the moral sense which seems to be absent in the animals? It may be admitted that genes are selfish but this does not explain people like Mother Teresa. Challenged about her, Dawkins said that her behaviour was motivated by a heavenly reward without any real concern for those whom she tended. But such a desire for heaven, which is debatable, does not help her genes which he regards as selfish. He thinks that altruistic genes are rare and act to eliminate themselves. What we know is that laws of nature have nothing to do with conscience or moral sense. If I jump from a high building the law of gravity will ensure that I fall to the ground but such a law has nothing to do with my intention to jump.

It is not necessary at this point to enter into the debate about gene therapy and, germline engineering, except to say that the former appears more acceptable than the latter.

The religious nature of the subjects

The religious view is that there is something wrong with us, traceable to the fall of mankind. Such a fall is questioned by science for death was present before Adam and Eve. Perhaps in Genesis it means spiritual death, separation from God, not physical. But if there was a state of original perfection or righteousness it conflicts with evolution: an evolving complexity from which human consciousness has gradually emerged. And can the tendency to sin be inherited? Modern genetics does accept the inheritance of characteristics but Judaism or Islam do not mention a fall, though the Hebrew scriptures have the evil imagination. Paul sees a connection between Adam and the human race which might be interpreted as meaning that sin is like a disease which has infected us all (Rom. 5:12). But many prefer to think that we repeat what Adam did so that each man is the Adam of his own soul.

To establish more of a connection it has been suggested that Adam the ancestor embodies, symbolises and represents the whole group of his descendants. In the creation story he is 'the man' confirmed by Paul (1 Cor 15:22, Rom. 5:12–21). He is the representative figure and typical man, everyman, whose story is re-enacted by each of us and he is mankind, the corporate entity to which all of us belong. It is the natural solidarity argument and we are related by inheritance, imitation and involvement. In contrast there is now the possibility of a new solidarity of grace in Christ.[19] We might even suggest a sin gene or a sin meme but this is basing it on the physical, whereas the result appears (as we have noted) to be spiritual death or separation from God. In any case, the story of the fall was based on the empirical: the Jews saw clearly, as Christians do, that there is a corruption in humanity and told the story to account for it.

Science has shown that a creature has emerged with a self-consciousness which has given him power over his environment and enabled him to survive. Religion makes him aware of failure to fulfil his potentialities and this characteristic has perpetuated itself by transmission within his culture. There is a social bias that thwarts the divine purpose, hence sin is also cultural and corporate and consequent upon origins. We demand power and are self-centred, seeking to become like gods (Gen. 3:5). In a certain sense we can associate the fall with a presumed period of historical time when humans became self-conscious and able to choose. It was then possible for events to occur contrary to God's creative purpose.[20] Man was created for glory not for sin, with good potentialities though he failed to grasp them.

Thus the testing of the experiment began when self-consciousness emerged due to a complex brain. Humans opposed the rules of the creator, choosing to follow their inherited instincts. We do not speak of a fall before this, since we cannot blame animals from following their genetically influenced behaviour. But why should we go against it? Because we are like God (*imago dei*) and have a relationship with him. Russell Stannard thinks that Jung's idea of the collective unconscious is helpful. It is the common thought patterns of humans arising from archetypes or imprints. Such unseen organising principles govern our conscious thoughts and attitudes and there is the shadow archetype which we refuse to acknowledge. That is the dark side of personality but there is also the religious archetype which appears everywhere. Dawkins calls it a virus which is self-replicating and means that most people have been sick! But on this basis could we not also say that atheism is a virus? Freud saw religion as a projection, Jung saw it as beneficial, hence their different interpretations, but the imprint implies an imprinter and Jung did believe in God.[21]

The religions recognise the power of evil and sin and in Christianity there is the atonement of Christ to release mankind from such a condition (Mt. 20:28). Christ passed the test which the first Adam failed to do, but creation and salvation have to be kept together as the divine intention from the beginning. Otherwise we have a record of failed experiments and Jesus coming to repair the creation. The idea of the two Adams keeps creation and redemption together. The purpose of creation is to bring another personal creature into fellowship with God and it has been accomplished in Christ. The doctrine of election points forward to this redemption and perhaps the best statement of it is in the writings of Karl Barth. He contends that all have been elected in Christ with the world as a theatre of God's love but some may opt out of it. God has elected Jesus to bring into being a new community and none is excluded except by their free choice. Christ is both elected and rejected but in him God has elected humanity as his covenant partner.

Hick, however, is not happy about the traditional doctrine of creation and fall and prefers to follow Irenaeus, who taught that Adam and Eve were like children and called to grow into perfection. Irenaeus wrote, in *Against the Heresies*, 'that man should in the first instance be created; and having been created should receive growth; and having been strengthened should abound; and having abounded should recover; and having recovered should be glorified; and being glorified should see his Lord' (4.38.3). Hick argues that God created finite personal beings at an epistemic distance from him, a distance of knowledge, hence they were under no pressure to believe in him. If he had not done this it would have affected their freedom and autonomy. The distance meant that, while conscious of their natural environment and the struggle for existence, they were only dimly aware of God, being childlike and immature and needed to exercise faith. But unlike the animals man has a religious bias in his nature.[22]

The accounts in scripture indicate a kind of dual causation. Characters act independently of God, yet the result turns out differently from what they expect because of the action of God. Joseph recognises what his brothers have done to him and that they meant it for evil but God makes it result in good (Gen. 50:20) or in the case of Jesus: 'The Son of man goes the way that is written of him, but alas for the man by whom the Son of man is betrayed' (Mk 14:21). From the theological point of view God is in continual interaction with the experiment and can change his mind concerning what he will do (Isa. 38:1), as in the decision not to destroy Nineveh, and he repents that he made Saul king (1 Sam. 15:11). But God does not make mistakes and then have to change his plans for his appointment of Saul was conditional on his loyalty. Saul refused to obey the word of the Lord and was rejected.

The materialist does not recognise the spiritual, contending that we are nothing but a combination of chemicals: carbon, oxygen, nitrogen, sulphur and phosphorus. But this view neglects any kind of personal explanation and does not take into account the higher levels of being. The reduction to the lowest level loses sight of how things are arranged in a certain order which conveys meaning in a way that the elements by themselves could not have. For example, when hydrogen and oxygen are combined it produces a new property: water. In our case the highest level is the personal but on lower levels there are cells, atoms, molecules, protons and electrons.

Matter has the potential to evolve into persons and it is a more intelligent process if there is a cosmic scientist engaged in design, but the emergence of self-consciousness involved the risk of man seeking to be independent of him. The complexity in such development would be unpredictable which is well put by Brian Wren:

> 'Are you the gambler-God, spinning the wheel of creation,
> Giving it randomness, willing to be surprised,
> Taking a million chances, hopeful, agonised,
> Greeting our stumbling faith with celebration'?[23]

There have been many views of the the image of God, which has been defaced by sin but renewed through Jesus Christ. The scripture by means of symbol and metaphor conveys the dilemma of the human condition and Christ's atonement: financial, legal, military and sacrificial. A picture then emerges not of a detached scientist but one who loves and sacrifices so it would appear that these qualities are inherent in God and not unforeseen, for there was the Lamb slain before the foundation of the world (Rev. 13:8).

Free will

Are we really free creatures who can choose between good and evil? We would argue that the experiment needs the freedom of the subjects, with the cosmic scientist allowing us to develop without coercion. Free will has been mentioned in connection with genes. The philosopher who is a compatibilist works within a framework of determinism but distinguishes between things that happen and those that are the result of our desires. The desires are there and she is not interested in where they come from but how they affect her choice. Thus freedom is acceptable but in a weak sense. She asks: What are we really like? How do we view ourselves, our desires and intentions, and how do others view us? There is the moral sense: I ought to do it and there is no point in having this feeling if I cannot perform the action. As Kant said: 'I ought therefore I can.'

To discuss where we got this moral sense, either from revelation or evolution, is to go back to origins. But there are also future ideals, aims and goals, and the future can have as big an effect as the past. If the determinist says they are also determined how is it that I reject some and choose others: is that also determined by my past? We could say this about a thing like a clock but hardly about an organism that is growing and developing. Character grows in stability and controls our thoughts and actions but our characters can change, unlike that of God, because we are in an environment which has evil as well as good. Choice and responsibility will affect our characters, but we cannot punish the offender unless there is free will.

Some appeal to quantum mechanics, where there is a lack of predictability and cause. But if there are uncaused events, I cannot be held responsible for them. The scientist works in a framework of order for if everything was haphazard he could not make progress. Newton's laws work with ordinary velocity but Einstein showed this was not so at high speeds. The matter seemed to be settled, but then came quantum mechanics and the uncertainty principle. We still have regularity but much more openness and flexibility and the humble feeling that there is still a lot more to learn. Hence the search for a unified theory. Hawking seems optimistic but even if he or someone else gets there in the future the history of science indicates that there will still be much to learn about this universe and its operation.

Soul

There is a tendency now not to use the term 'soul', though *psyche* can be translated in this way. But the NEB translates Mk 8.36–7 using the term 'true self'. Other words such as *bios* and *zoe* mean life. Keith Ward, however, uses mind and soul in an interchangeable way. The goal of evolution is self-conscious purposive moral beings. The mind or soul is dependent on the brain but in perceiving and understanding goes beyond it and appears to be an organising principle that interprets things and gives a sense of continuity throughout life. It is like a compact disc storing information. Some theists are willing to accept this, but the disc is a material thing, hence Ward points out that the discs must be read by someone. The reader is the soul which works with the brain but directs and coordinates. It is not just a complex pattern of physical activity for it responds and acts, so we are embodied souls with personality. The hope of resurrection is for a new body that will express the true nature of the soul.[24]

In conclusion we can say that the varied ways of explaining the entry of sin into the world arose out of the evident corruption which we see constantly. In the words of the Apostle who was dismayed at such a fallen creation it was like a woman in childbirth groaning for the birth of a child. But when a child is born she rejoices and forgets her pain. In the birth of the Christ-child a new hope dawned for the world, for there was the belief that with his resurrection there will be a new body for humanity and with the coming of the kingdom of God a new heaven and earth (Rom. 8:22ff.; Rev. 21). It will be discussed in subsequent chapters.

Evolution could be described as a vast experiment, a dynamic process reflecting a continuous and flexible creativity in the process. The divine scientist enters the experiment and seeks to persuade, not compelling his subjects but allowing freedom and spontaneity in nature. Such action will be discussed later, but now we turn to consider the greatest problem for the theist: the existence of evil.

6
The Test

The previous chapter raised problems for a divine designer but evolution also questioned the goodness of God in that there was much competition, waste and needless suffering. Darwin's loss of faith was basically concerned with this problem and today it still remains a major obstacle to belief.

In this chapter it is important to see if the model of God as the cosmic scientist helps to resolve the presence of evil in the divine laboratory. The model is based on the cosmic order shown in the regularity and intelligibility of the universe, allowing the scientist to perceive patterns which need explanation. The amazing thing is that our minds are tuned to discern them and describe the laws. It is true of course that knowledge of the laws changes and that they imperfectly reflect the regularities.

Order reflects design: but what about disorder? God as creator and designer would be omnipotent. Why then does he not abolish the evil disorder? Does this not question his goodness? If he is good, he should deal with evil but does not; is he then powerless? It is the dilemma presented by David Hume but a lot will depend on how we define omnipotence.

We mentioned earlier that God can do everything that is in accord with his goodness but he cannot do the logically impossible, for example, square a circle. He cannot give us freedom and at the same time save us from doing wrong. Peter Vardy thinks that Kenny's definition is a good one: the possession of all logically possible powers which it is possible for a being with the attributes of God to possess. But our freedom has placed a restriction on his power to intervene and we cannot judge his goodness by some independent standard. Yet it will be greatly superior to what we consider it to be.[1]

Origin of evil

How did evil originate? We see in the world both moral and natural evil, with the innocent suffering, and disasters such as earthquakes and volcanic eruptions. Moral evil is widespread: crime, sexual assaults, hatred, and violence. How did this moral corruption originate? Is there a force for evil that drives mankind away from God and towards corruption? Stories in various scriptures postulate evil forces or fallen angels (Gen. 6:1–8) and, after the exile (Wisdom 2:24; compare Rev. 12:9; 20), the serpent in the garden was identified with satan. The fallen angels' story draws on pagan mythology where giants, Greek Titans, were held to be offspring of divine/human intercourse. But the symbolic significance is that evil is cosmic affecting the whole creation (Rom. 8.20–2). It is a power which works through tyrants, dictators and systems, but Christ has already won the initial victory over them. John Hick thinks that we cannot discard the possibility of evil spiritual beings but today we think in terms of psychological forces such as mental complexes and psychoses. That is to see evil as internal, though the scriptures speak of an external force and personify satan (John 8:44; Lk. 22:31; Hebrews 2:14ff; 1 Pt. 5:8; Eph. 6:11–12; Rev. 12:9).

Christ speaks of him as the prince of this world (John 12:31–2; 14:30) and Isaiah refers to the bright and morning star, falling from heaven (14:12–15). Commentators think that Isaiah is referring to the king of Babylon[2] but Paul is sure that there are powers, authorities, thrones, lordships, and angels (Rom. 8:38) and in the letter to the Ephesians (6:12) they are the cosmic potentates of this dark world and superhuman forces of evil in the heavens. He is following the

Jewish belief that there were many ranks of angels and some were hostile to the creation of humanity. It is true that the world of the first century was peopled with demons which nowadays the mass media portray as complexes in the depths of our unconscious. But Plantinga says that the existence of the devil is a possibilty and refers to the temptations of Jesus and the force of evil on earth.[3]

Evil was in the world prior to the human fall so can we equate it with the *tobu wabohu* which is the opposite of order and which appears to exist prior to the creative act? This formless void or chaos could not be the creative act of God since his judgement on the work of creation is that it is good. The doctrine of creation out of nothing (*creatio ex nihilo*) however did not appear until 2 Maccabees 7:28 and did not take firm root until the second century in opposition to any idea of the eternity of matter.[4] The point of the doctrine is that the world is not co-eternal with God but it could be contended that the *tobu wabohu* was caused by a premundane fall. What is clear from evolution is that evil was in the world before humanity, with wastage, suffering, natural disasters, and the wiping out of many species including the dinosaurs.

Testing

As we saw, the garden of Eden did not symbolise a paradise, for there was evil there which implied a test for the first humans. They may have been immature creatures belonging to the childhood of the race, as Hick believes, who would develop through the test of suffering. Christian teaching has stressed the fall but neither Judaism nor Islam have the doctrine. Testing continues with Abraham being called to offer the sacrifice of his son and Jacob was changed into Israel through his trials. Joseph was hurt and hated by his brothers, and Job suffered grievously. No prophet in Israel was exempt from it. Jesus was subjected to tests and the temptations centred around how he would use his unique powers. There would have been no point in these temptations if he had not had the power to turn stones into bread or jump from the temple. Through his ministry the tests continued, for he accused Peter of being used by satan (Mt. 16:20–3).

Temptation in the New Testament means any testing situation which examines a person's character. Those who heard Jesus were tested by their reaction to his teaching and the values he put foward (Lk. 6:2–26). He did not hide the fact that he was to suffer and the same was asked of them (Lk. 9:57–62). In the event many turned away from following him. Peter failed the test by his denial (Mt. 26:69f.) but repented and was reinstated and Jesus was tested to turn away from the will of God in the Garden of Gethsemane (Mt. 26:36–46). The Lord's prayer (Mt. 6:9–13; Lk. 11:2–4) includes the clause, 'do not bring us to the time of trial, but rescue us from the evil one'.[5]

The idea of testing occurs in religions other than Christianity. The Buddha was tempted not to embark on his mission and in Hinduism, Arjuna, according to the *Bhagavada Gita*, is challenged by the incarnation of Vishnu, Krishna, to go to war and kill his relatives. He refuses but Krishna says that he will not be able to kill their souls and therefore he must do his duty as warrior. A somewhat dangerous justification for war but Gandhi, the pacifist, interpreted the story in the spiritual sense that we are called upon to fight evil internally.

It would seem that the purpose of God, as the divine scientist, is to create a testing ground, the laboratory of the world, for us to achieve the values which he wants. It is the means of achieving beauty, goodness and truth, whereas a paradise would not encourage us to seek these things. The main objection is that the price is too high. No one wants suffering but the Psalmist said that it was good for him to have been in trouble (119:71). Perhaps he said it because he had learned more from suffering than from pleasure. The case of Job shows God allowing satan to afflict him in order that his faith would be tested. But can the means justify the end? The same could be asked concerning the cross of Christ who, according to the record, saw the cross as the means of drawing all to him.

We are contending that the idea of testing might be one answer to suffering but it might be objected that God could have made the world a better place so that such suffering was not required. Why not create a paradise where there would be no struggle or danger? Such a place as we have said would not produce courage or endurance or patience and we would be programmed to choose the good. Free will means that God has granted us space and does not coerce us to do what he wants. In our final chapter we will see that God intends to create a new heaven and earth but it will be for those who have passed the tests.

The absence of God

What makes the situation more difficult is the absence of God when the testing is imposed. The Psalmist was asked the same question as those who suffered in the Nazi concentration camps: Where is your God? He asks, 'Why have you forgotten me? Why do you hide yourself in times of trouble? (Ps. 10:1) The feeling of being forsaken is also heard in the cry of Christ from the cross (Mk 15:33–4). Is it just for the innocent to suffer and the guilty escape? The Psalmist cries: 'All in vain have I kept my heart clean.' But then he goes to the Temple and the truth dawns that evil is only temporary. As the apostle Paul was to write later: 'For this slight momentary affliction is preparing us for an eternal weight of glory beyond all measure (2 Cor. 4:16f). He lists his afflictions: calamities, beatings, imprisonments, riots, labours, sleepless nights, hunger. ... yet suffering is viewed as constructive with Paul rejoicing in it and over it (Col. 1:24). And in Hebrews it is seen as a discipline (12:7, 11) which yields the peaceful fruit of righteousness. Paul had 'a thorn in the flesh, a messenger of satan' and he prayed for its removal but God refused and gave him grace to bear it (2 Cor. 12:7–10).

Jesus himself felt deserted by God in his suffering (Mt. 27:46) and it may be that he had Psalm 22 in mind when he uttered the cry, 'My God, my God, why hast thou forsaken me' which is the first verse of the Psalm. But it ends in triumph, praising the control of God over all (25–31). According to Paul, Jesus was bearing the sin of the world, which meant separation from God, but it also might mean that he was plumbing the uttermost depths of what we feel when suffering strikes. His final word was, 'It is finished' (*tetelestai*), the shout of the victor, of the one who has completed his task. It is an encouragement to hold on to faith even in the darkest moments of despair. The innocent do suffer but good can come out of evil and there will be ultimate justification.

The fall

We mentioned the fall doctrine in the last chapter but the question arises: if God knows everything, how did he allow sin, which was to bring so much suffering, to happen? The suggested answer is free will but Augustine and Calvin produced a doctrine of predestination that God knew and elected some but not others. How could God have withheld his grace from some in this way? A better understanding is found in Sikhism where predestination means permitting rather than determining. In any case we have argued for the modification of the divine attributes in the light of creation. John Hick rejects the fall and contends that the myth is an analysis of man's state as it has always been,[6] but he does not reject the demonic in the light of the sadistic cruelty of our age. It has perverted and corrupted human beings who appear to have followed Milton's satan: 'It is better to reign in hell than serve in heaven.'

Perhaps evil exists within the divine good purpose. Augustine said: 'God judged it better to bring good out of evil than to suffer no evil to exist.' In the Augustinian type of theodicy the fall plays a central role in the divine plan and the origin of evil is an original conscious turning away from God. But Hick contends that in order to turn away there must have been some flaw either in them or their environment.[7] This flaw in creation must then go back to the creator. He prefers, as we noted, the Irenaean (140–202 CE) theodicy, in which, while the fall is not denied in all its forms, the doctrine is played down. The loss of the righteousness of man, the perfection of his world, and inherited sinfulness as a result of the fall are rejected.[8]

Hick is aware of the problems in the Irenaean position and lists them: God has ultimate responsibility for all existence including evil and that is hard to reconcile with a recognition of the demonic character of much evil both moral and natural. The affirmation of universal salvation is not easy to reconcile with our freedom and responsibility. But it seems to Hick that these difficulties are of less moment than those attached to the alternative Augustinian doctrines of the pre-historic fall of men or angels and eternal hell for a proportion of God's creatures.[9] Both positions, however, hold that the world was created by a good God who continues to sustain it. Irenaeus believed that God uses evil for the good purpose of morally developing us.

We need not enter into this debate, for whatever view is held both accept that the test of suffering develops character and that evil is some kind of disorder which entered into creation. Evil is not negative non-being, as in some past philosophical discussions, but a reality that must be defeated. In his ministry Jesus faced that reality as he dealt with the sick, the diseased, the blind, and sinners who sought forgiveness. God, he taught, was a Father who cared and comforted them. If we believe this then God is not only the scientist who has given laws to the universe but who seeks in a personal way to suffer with us. But in seeing evil as a reality and not simply the absence of good we need to remember that it is not substantial but, as Aquinas said, a perversion of the will.[10]

Evolution encourages us to believe that pain is inevitable in any creature who gains information from its environment and enables it to avoid dangers. New forms of matter arise only through the destruction or death of the old, so evolution justifies the predation of one animal on another. Old patterns must give way to new and the death of some ensures food for new arrivals. Nature selects the fittest, leading to change and improvement of the genes. If there were no conflict then overpopulation would be the consequence. Morally we find it difficult to accept but it is there in nature and in ourselves, namely the survival of the fittest. We say that it ought not to be like this so we care for the disabled and the old, and religion insists that the final test will be how we have done this.

But if we must experience pain to gain good character, need there be so much of it? We care deeply about the victims of famine, earthquakes, typhoons, volcanic eruptions, cancer, polio, meningitis. And how does the suffering of an innocent child born with a disability develop character? I recall one of my disabled students telling me that according to certain liberal thinking she should have been terminated after birth. But her parents cared for her and she was continuing to work for a university degree. We are always amazed at what those confined to wheelchairs can do, and the loving work of carers on their behalf. Looking after such people draws out great qualities of patience and endurance.

We might also argue that these events and births are exceptional, for the majority of humanity enjoys the order and regularity of the universe, undisturbed by earthshaking disasters, and continue to have healthy children. There is the beauty of the world, the joy of living, and when disaster arises we rush to help those in need and praise the extraordinary efforts made by surgeons to save life. What is odd is that, despite the predictions about the moving plates in the earth's surface, people continue to live in earthquake zones. It is foolhardy to do so but many prefer the beauty of such places and take the risk. We need to remember too that if the creator has given a freedom to nature to make itself then he is permitting these things to happen but he does not intend them. Creation as it now is, is not the divine intention but it is hoped that it will be fulfilled in time.

But why not a greater degree of stability which would rule out natural evil? We noted in Chapter 4 how finely tuned the world was, yet disorder persists. Since this is so, wherever it came from, how could God intervene in some cases and not in others? If he was continually interfering it would disturb the normal course of life. And if there was no disorder how would he test our love and obedience? Nevertheless, some are embittered by a suffering which does not appear constructive. There seems to be a choice between the pessimism of

Ecclesiastes: 'Vanity of vanity, all is vanity' (1:2) and the apostle Paul's verdict, 'All things work together for good to them that love God' (Rom. 8:28). Yet he also recognised the mystery of evil and said that now we see through a glass darkly (1 Cor. 13:12, 13). What we need are faith, hope and love.

One of the difficulties in traditional theology was the belief that God had specially created everything. How then did we explain their imperfections? But if evolution is God's way of creation, with natural selection operating, then it helped to solve the problem. Defective design in any creature can be explained by evolution slowly modifying the existing structure so that defects, deformities and dysfunctions are due to natural causes.[11] It would seem that God did not plan the details of evolution from the molecules of DNA to complex organisations but controlled the whole. We have a general blueprint not a determined plan, a system set up to give rise to organised complexity, and we do not know all the causal factors. But in the evolutionary process there have been many mistakes and dead ends.

The Holocaust

Such speculations relieve God of some of the responsibility but some continue to doubt. Jung reacted against Freud's view of religion as an illusion but he saw a shadow in God. There is a paradox here reflected in the scripture where God is responsible for evil (Amos 3:6), but nevertheless hates it (Isa. 13:11, Dt. 19:19–20, Jeremiah 4:4). Evil is clearly the enemy of God in the Gospels and Epistles and Jesus wrestled with it thoughout his life. The Hebrew, however, was inclined to trace everything to God and did not seek for natural causes. But if we go on to say he uses evil for a good purpose we are faced with explaining such awful events as the Holocaust. The old question of the psalmist as we noted returns: Where was God? We cannot say that God intended it to happen, so the question changes: Where was humanity? How could a civilised race perpetuate such crimes not only on the Jews but the Russians? Can it only be explained by the corruption of humanity or is there, as we have noticed, the possibility of evil forces which can take control of humanity? But why did God not intervene to stop such a tragedy?

Some think there is a hidden God as well as one who reveals himself in the bible.[12] Other Jewish thinkers believe in contraction: God had to contract or withdraw into himself to make room for the universe, so he is hidden as well as revealed. They think of God paradoxically. Scripture teaches that Israel was elected as a priest people which means that it must give its life blood as a ransom for humanity. It is ever bringing new sacrifices as shown by the symbol of the lamb. Others reject election saying that it implies favouritism but the biblical account of the covenant makes clear that it is for service not privilege or superiority over other people.[13]

Suffering is a mark of being chosen. The Holocaust cannot be for sin that the Jews committed and in any case is there any sin in the world deserving of such punishment? And is there any punishment in the world capable of compensating for the crimes committed against the Jews? If God has veiled his face in the light of such suffering when will he unveil? What are the limits of God's forbearance? Yossel Rakover asks these questions and goes on to plead with God not to put the rope under too much strain or it will snap. Some have already turned away from God, unable to explain or bear the suffering. Rakover prays for mercy to be granted to them. But his own faith climbs the heights: 'I tell you this because I do believe in You, because I believe in You more strongly than ever, because now I know that You are my Lord, because You cannot possibly be after all the God of those whose deeds are the most horrible expression of ungodliness.'[14]

There is no consensus of opinion about the Holocaust since there is no rational explanation. Many felt that they must keep the faith and look for a life after death to rectify the scales of justice. It is also held that good can emerge out of evil, for the Holocaust spurred the Jews to establish the state of Israel where they would be free from tyranny. The victories that Israel gained in 1967 and 1990 over their enemies encouraged them to believe that God was now acting on their behalf. But others continued to argue with God about it and some lost their faith.

Another explanation is that God was affected by what happened to the Jews so that in their tragic history of suffering he suffered with them. It departs from the traditional view of divine impassibility. If God cannot be localised in a Temple then when the Jews went into exile in Babylon he went with them. The same occurred when their Temple was destroyed by the Romans in 70 CE. The old story of Daniel in the lions' den is intended to show that there is a divine presence in suffering and such thinking can be extended to the presence of God in the Nazi gas chambers. It is a mystical interpretation and often associated with the Kabbalists but it testifies to an interaction of God with his people and reacts against any deistic view. However, when the Kabbalah saw good deeds helping to restore the divine unity which had been affected by sin, it was condemned as a heresy.

What the Holocaust shows is that suffering happens in our world which God does not intend. The brothers of Joseph intended to get rid of him and committed him to slavery in Egypt but God intended something different for him and brought it to pass. Much suffering is caused by free will, which can be a dangerous weapon in our hands. God wanted free creatures who would respond to his love but such freedom entails wrong choices which he does not want. The price of such freedom is often suffering.

God, it would seem, has voluntarily limited his power by granting us freedom. If he intervened he would be infringing that freedom and if he interfered in nature with its earthquakes and disasters and cancers, he would not be allowing a world to make itself. The concept of chance in evolution could imply that he has given us and the world room to manoeuvre. A mechanical picture of the world no longer prevails but there is now seen to be a variety of causes, some of which are more easily detectable than others. But such freedom does not mean that God has lost control of the world. In Revelation 6, the four horsemen representing invasion, rebellion, famine and pestilence are let loose by the crucified lamb who is the ruler of history and allowed to have their way. But it is only for a little while. The initial victory over them has been accomplished by the cross and resurrection.

Other reactions

Evil can also arise because of incompetence. Jane Hawking asserts that the illness of her husband Stephen might have been due to a non-sterile smallpox vaccination given in the early 1960s. But she thinks that though he is not a believer his fight against it might be divinely inspired. God does not intervene in suffering because of our free will but he does give help and love. She believes that in parting from her husband she received help and her personal relationship with God was deepened. It is clear that she relied very much on her faith in those testing days and concludes that 'however far-reaching man's intellectual achievements and however advanced his knowledge of creation, without faith and a sense of his own spirituality, there is only isolation and despair and man really is a lost cause'.[15]

Just as Job is not given an answer except shown the majesty and care of God in creation, a final answer to the suffering of the Jews is not explained by any of the above responses. But the history of Israel shows that God never deserted them and was always present. Can we doubt then that God was with them in the gas chambers? The Jew has every right to feel bitter about his suffering but some said that love is better than hatred and are determined never to become like those monsters who tortured them in the concentration camps. As Itka Zygmuntowicz says of her suffering: 'There are those who claim that love is blind, but it seems to me that hatred is blind. Love builds bridges of communication, and hatred builds walls of isolation. Hatred divides us and destroys us, and love protects us and unites us.'[16] Kant, contemplating so much undeserved suffering in the world, said that there must be a God who will redress the balance in another world. We can say that good character can emerge from the discipline of pain but it does not account for all suffering. Others cling to the hope that we cannot understand now but will know hereafter.

The story of Job shows the tempter demanding a test of his faith in God. he insinuates that Job is trusting in God for gain. The test concerns the motivation of his faith. Job thought some suffering was to be expected but not the extent of his, and he is grieved that his friends think he is being punished for his sin. His wife turns against him, telling him to curse God and die, but he first of all resigns himself to it, saying 'the Lord gave and the Lord has taken away, blessed be the name of the Lord' (1. 21) and his faith rises magnificently: 'Though he slay me yet will I trust him' (Job 13:15).

In the final chapters Job sees God opening up the panorama and becomes aware of his humble place in it. Pain cannot be abolished but there is a whole creation to share it and compassion has to be learned. His repentance can be translated, 'I am comforted concerning dust and ashes, comforted concerning my finite humanity' (42:6). He lived on but his understanding of life was altered. When God showed him his power, purpose, greatness and care for all creatures Job found peace.

The message seems to be that God is telling Job that there is much Job does not know or and cannot do, for he is ignorant of the laws of heaven and an understanding of the ways or thoughts of God. All our models of God are inadequate but interestingly Job is directed to gaze at the heavens (Chapters 38 and 39) and what is going on in nature, something which has always had significance for the astronomer, the physicist and the biologist. Is this not another pointer to the importance of the scientist and natural theology? And does it help in showing that God is a scientist?

Turning to Indian philosophy we have the concept of *karma* which is the reason for the suffering caused by human wrongdoing. *Karma* is considered to be a natural law just like gravitation. But why does Brahman permit *karma* to exist? Because he wants souls to work out their *karma*, which involves suffering as a discipline. In that sense the law is part of Brahman's nature and inequalities are explained as the result of sin in a previous life. The doctrine rests on the validity of the transmigration of eternal souls, 'the wheel of life' (*samsara*), otherwise the question arises: When does the responsibility for what one does first occur? With *karma* there seems to be a mixture of predestination and free will. The Indian tradition does not speak of God as judge who will acquit or condemn at the end but *karma* might be seen as similar to the principle that what we sow we will reap.

In the teaching of Jesus there is a recognition that sin brings suffering (Mk 2:5) but it was not always the case. He pointed to the death of Galileans caused by Pilate or accidental deaths caused by the collapse of a tower (Lk. 13:1–5), and that it was a waste of time to ask about the transmission of sins of parents to children when good could be done. The question arose in connection with the man born blind where they asked him if the man had sinned, or his parents. The answer of Jesus implied that it was time to cease from such arguments and seize the opportunity of restoring his sight (John 9:1–3).

Evil showed itself in the temptations and the condition of the sick and the opposition which he encountered. Continually during his ministry he sought to bring good out of evil and in death he turned the cross, which was a thing of shame and curse, into a symbol of glory. The cross was not the cause of good but the occasion of it. Evil is overcome by love and sacrifice not by force. The picture of Christ is that of prophet, priest and king, with the priest offering himself for the sin of the world (Heb. 9:14) so sacrifice is the source of new life.

Undeserved suffering can have a redemptive effect on others: the suffering servant of Isaiah 53, and the thief on the cross. Without suffering how could there be such moral courage? It evokes compassion and love. Creation is continuous in the sense that we are being made spiritually (Eph. 2:21; Col. 2:19). The divine laboratory is the vale of soul-making, not some kind of playground. It is, as Hick says, a leading from *bios* or biological life to *zoe* or personal life of eternal worth. Thus there is a teleological aim.[17]

Christ as the suffering scientist

For the Christian, Christ is identified with the suffering servant of Isaiah 53, but in a scientific world we could see him as the suffering scientist. The scientist in our culture is a symbol of one who battles with disease and evil and can involve himself in risk. Scientists sometimes have to endure suffering personally in order to find cures for diseases. Of course the physical sciences deal with objects but in the social sciences, psychology, and medicine, that is not the case. Here a battle goes on against mental and physical diseases and is analogous with Christ's work of healing the sick, forgiving the sinner, and dealing with the mentally ill.

In his battle against evil he goes all the way to the cross and endures the worst that it can do. Likewise the scientist has often to make agonising decisions and emerges from his laboratory exhausted. In the case of Marie and Pierre Curie, the discoverers of radium, they believed there existed some element which no one had yet isolated or seen. They conducted a series of experiments with pitchblende, trying to isolate this new element but knowing little about it were unaware of the risks. But at last they succeeded, for one night they saw the glowing of radium in the darkness. But a price was paid for the discovery. Marie Curie carried in her twisted and deformed hands the evidence of the strange power of the radium to inflict suffering on those who handled it and she died prematurely.[18]

Newton neglected his health in writing the *Mathematical Principles of Natural Philosophy* and the results of twenty years of work in optics were lost in a fire. In order to understand the nature of light he stared at the sun until he almost went blind and stuck blunt needles behind his eyeballs. His sight only properly returned after a period in a dark room. He wrote extensively on alchemy and tasted a variety of toxic substances which may have unbalanced his mind.

Lavoisier, the great chemist, was accused of stealing other men's ideas, thrown into prison in 1793, and sentenced to death. When the court was told that, in the opinion of most of the scientists of Europe, citizen Lavoisier occupied a distinguished place among those who brought honour to France, the President commented that the Republic had no need of men of science. In medical science, Lister had to fight for antiseptics in surgical operations and Simpson battled against opposition in the use of chloroform.

In the last century 200,000 people died as a result of the atomic bomb being dropped on Hiroshima and Nagasaki. J. Robert Oppenheimer in 1949 struggled to get the military to agree to demonstrate its power with Japanese observers present. It was opposed on the grounds that it was moral for the Americans to do what the Japanese had done to them at Pearl Harbor. Oppenheimer refused to work on the hydrogen bomb and in 1954 his security clearance was revoked and he was forced from public office. He died of cancer in 1967.

Alan Walker also illustrates the point with stories of scientists risking their lives for the work of another. During the First World War many soldiers suffered wounds infected with tetanus or gas gangrene and a doctor named Taylor suspected that the condition of the wound was caused by a specific bacillus. He tried many experiments but needed to inoculate another person with the remedy to see if it worked. A nurse called Mary David inoculated herself with a preparation of the bacillus and when Taylor was told he immediately injected the chemical remedy which he had prepared and waited for some time while the life of the nurse trembled in the balance. But to his joy she was eventually pronounced out of danger. By taking the risk of losing her life she was the instrument of saving others. The scientist has to battle against mistrust of new drugs and the danger that some people see in such advances.[19]

The cosmic scientist is engaged in warfare against every kind of evil. T.S. Eliot uses the image of God as 'the wounded surgeon' and others speak of the vulnerable God. The cosmic dimension of such a conflict emerges in the writing of St Paul and the victory of the cosmic scientist over it (Col. 2:15). Paul, as we noted above, when confronted with the mystery of suffering, mentions three instruments to deal with it: faith, hope and love. In Hebrews the importance of a faith relationship with God is emphasised. The test is to have faith in God despite circumstances, as in the case of Abraham, and without it we cannot please God. Habakkuk said that the righteous live by faith (2:4) which is the conviction of things not seen (Heb. 11:1), and necessary if miracles are to be performed (Mt. 8:5–13; 9:1–8; 9:20–2). According to James only true faith and commitment will resist the devil and enable us to have patient endurance when suffering comes (4:7). Faith is allied to hope, for we know that our suffering is only temporary and there is a possibility of injustice being put right after death. In the present life love is essential and we have seen a demonstration of it when thinking of the reaction of many to the hatred of enemies.

It is reasonable to argue that God does not intervene when we suffer because it would coerce faith. Pain acts as a warning signal and suffering sometimes results from loving someone. As Paul says: 'God gives proof of his love towards us in that while we were sinners Christ died for us' (Rom. 5:8). It is the cross that reveals the extent of the suffering that love will go to, and it tests our faith, resulting in it being refined like gold (1 Pt 1:7). Stannard refers to the making of a pot. The clay is subjected to extreme heat otherwise it will remain soft clay. The test of fire must be passed for it to be transformed into what is useful and not all clay pots survive the test, for some reveal a flaw. According to scripture Jesus was crowned with glory because he was willing to suffer even unto death and deliver us from its bondage and fear (Heb. 2:15). He was tempted and tested and is able to help us to bear our suffering (Heb. 2:18). God vindicated him and will vindicate others. In any case suffering is limited but an afterlife where God will wipe away all tears would be infinite.[20]

In conclusion we can say that testing is one of the ways to understand suffering but it is only one possible solution and it is difficult to apply to some of the horrors of modern life. We are faced with a mystery, as the apostle Paul says, and while we cannot explain it we can take his advice to use faith, hope and love in dealing with it. But the question continues to be raised: Why does God not intervene to stop it and the usual answer of our freedom does not always satisfy. It is necessary then to consider God's action in the world, which will be the theme of our next chapter.

7
The Action of the Divine Scientist

In this chapter we concentrate on the action of God in the world, taking into account the views of scientists and theologians.

It is difficult to involve a totally transcendent God in events but the other extreme of immanence equates him with the world. How can God be both in the experiment and yet beyond it? One suggestion is panentheism, which is suggested by both Ian Barbour and Arthur Peacocke. But John Polkinghorne does not like it and is joined by Keith Ward. Their views are discussed in the context of the current scientific view of time and space.

Ian Barbour, a process theologian, thinks God is the great persuader who influences us in many ways but the picture of God that he presents seems to envisage him struggling in the process. A.R. Peacocke agrees without adopting process theology but Polkinghorne takes issue with him. He does not agree with Peacocke's view of Christ as an emergent in the evolutionary process or with Barbour thinking that he is a new stage in evolution. Polkinghorne contends that Jesus does more than inspire and inform: he redeems us from our sins.

Peacocke believes that God interacts with us by influencing the patterns of events. He says that God includes and penetrates the universe but his being is not exhausted by the universe. Polkinghorne does not like the language of panentheism, especially that of process thought, and pleads for simply balancing divine immanence and transcendence. He dismisses dualism of mind and body and advocates complementary aspects of what he calls 'one world-stuff'. A parallel here would be quantum theory with its complementary wave and particle.[1]

He sets aside Maurice Wiles' timeless single act of holding history in being, for it does not take into account the personal nature of God. And the idea that divine causality is hidden within secondary causes is too mysterious a gloss on natural processes. A better approach is to put foward a non-deterministic account of chaos theory which, like quantum theory, shows an openness of the world but is on the macroscopic level not the microscopic. This would enlarge our knowledge of causal principles, not set them aside. Such systems are holistic in the sense they cannot be isolated from their environment and they are not totally disordered but indicate a range of possibilities. The holistic means active information, top-down, which supplements the bottom-up causality of the energetic exchange between the parts.[2]

The implication is that mind interacts with matter and there might be holistic laws of nature driving evolution. What we need is the recognition of openness at the lower levels where there is quantum uncertainty. Polkinghorne says that Peacocke misunderstands him when he thinks that he is advocating God manipulating micro-events, for he is trying to see the divine action within the unpredictability of the physical process. He believes in agency in terms of active information but there is need to know how energy and information are related to each other. He says passive information, that is communication theory, is not the same as pattern-forming active information, which is how God acts.

There are principles of a holistic and pattern-forming kind, which is active information, and top-down causality influencing the parts. God interacts, without intervening, and shows his power in creation by acting through pure information input or the working of the Spirit. His power means that he permits all things, but not all happens in accordance with his will. Polkinghorne now believes that God allows divine special providence to act as a cause among causes, as we observe in the life and death of Jesus. There we have an interweaving of divine and human causes. With him God does new things. His resurrection was unexpected and did not correspond to the expectation of such an event at the end of all things.

The important point which Polkinghorne is making is this top-down causality: inputs of pattern formation through information. It takes into account the behaviour of the whole and could have a parallel with the action of mind and brain. But, just as top-down physical processes are difficult to discern, so is the action of God. The problem is, how can God act in a lawlike world? There has been a change in our attitudes to the laws, many of which were unchallenged for three hundred years but are now known not to operate at high speeds, according to relativity or in the area of the very small, which is understood by quantum theory. Is there a hierarchy of laws with the spiritual above the physical?

Information or communication theory is shown in computer networks and DNA in organisms. Polkinghorne fastens on this, seeing God as a communicator of information and can point to the divine logos as the communicator of the rational nature of the universe and its meaning. Peacocke too starts from top-down causality within the hieraracby of levels in an organism. He suggests that God is a top-down cause, a constraint or boundary condition, and the all-encompassing whole of which natural organisms are parts. As the whole he acts on the parts.[3]

Other theologians relying on indeterminacy inherent in nature work with 'bottom-up' causality where the parts at the lower level interact with one another. But this is confining him to the micro-level rather than the macro. It is more likely that God would act at the highest level: the mind of man. Mind is thought of as an organisational principle that emerges in organisms and results in pattern formation at the higher level. But his action is hidden within the unpredictable processes of the physical and the spiritual. Such action would be more like influence than determination. The position is analogous with human action. I can watch the physical movement of a person but why she is doing it requires a different explanation than observing her physical movements. Beliefs and intentions which are hidden from me come into play and her creativity and freedom must be taken into account.

Polkinghorne tries to combine the 'top-down' and 'bottom-up' causalities and thinks that a study of chaos would enlarge our knowledge of causal principles, not set them aside. Such systems are holistic in the sense they cannot be isolated from their environment and are not totally disordered but have a range of possibilities. He believes that the holistic means active information, top-down, which supplements the bottom-up causality of energetic exchange between the parts. But the problem is that the concept of 'information', unlike energy, needs further clarification. Paul Davies says that this is because it appears in many scientific fields. In relativity theory, it is information that is forbidden to travel faster than light and in quantum mechanics the state of the system is described by its maximum information content. In biology a gene is a set of instructions containing the information needed to execute some task. But is information always preserved in physical systems? What happens to the information in a star when it collapses to form a black hole? Davies thinks that in the future the clue to these and other problems may be found in quantum physics.[4]

Mind/brain

All these thinkers would hold that the main action of God would be at the higher level of an organism, namely the mind. We have argued that God has a scientific mind and has given us rationality, which Plato saw as the soul, and it is the reason why we can tune into the universe which he has created. But we need a brief comment on mind and brain which I have discussed in more detail elsewhere.[5] Traditionally mind or soul was regarded as the rational part and kept distinct from the body. But this dualism, though there are arguments for it, has fallen into disfavour and the same applies to the behaviourism put forward by B.F. Skinner and others. They argued that we do not possess non-physical minds which direct our behaviour but mental states are simply dispositions to behave in certain ways. Many today opt for the mind–body identity theory which contends that mental states are identical with bodily ones. It could mean that conscious events are just complex arrangements of low-level particles.

This position has difficulty with thoughts which defy such a mechanical operation and go far beyond it, as with Einstein's thought experiments. External behaviour will not tell others what is going on internally, and introspection must be taken into account. There does seem to be more in the mind than observable behaviour.

But if we accept that consciousness or mind evolves from matter then there is no dualism between a spiritual or mental entity and the body. It corresponds with the Semitic tradition of man as a psychosomatic unity and contends for the development of an awareness of being an 'I', a sense of freedom, and a feeling of transcendence over our bodies. There is an emergent quality in evolution, for an organism is not simply a static assembly of building blocks added together but a dynamic complexity in which new forms of organisation emerge. The wetness of water is not the property of its constituent atoms but arises out of higher levels of molecular organisation. A person is a distinctive emergent, a unity and a whole, which cannot be reduced or placed on the same level as his or her constituent parts.

Those who study the mind admit candidly that they do not know what it is but what it does. The brain is very complicated, with the forebrain, midbrain and hindbrain. The first divides into cerebral hemispheres with the left playing a special role for language so that any damage there can cause misunderstanding and defects in communication. It is more connected with the emotions than the right hemisphere which tends to the rational. But one area of the brain seems common to all the different tasks which it performs, namely area 46. It has been called the central executive, which is needed to coordinate thought and to switch back and forth between tasks. When we recognise an object we become conscious of what it is but we are not conscious of how we recognise it. Movement or motor action is thinking, remembering and planning. The mind appears to see the world and act voluntarily on it and gives us the sense of self.[6]

Knowing how the brain works is helpful but it does not explain the feeling that 'I am me' because many strands of social and interpersonal relations convey a sense of self. Philosophy makes an attempt, taking into account consciousness, self and person. Peter Strawson put forward a double aspect view of persons who have both states of consciousness and bodily characteristics. The concept of person unites these. There is one thing viewed from two different aspects. Hence we cannot reduce the person to mind, as the idealists try to do, or body and brain, as the materialist insists.

When high complexity is reached in an organism a non-physical property comes into operation, or a principle of organisation which enables us to understand, have purposes, and are conscious of moral responsibility. Since it is dependent on the brain it is not separated into body and mind. Computers have hardware and software but as we noted someone is needed to interpret the discs, which would be analogous to the soul or self. There is continual interaction between me and the computer and in the same way between soul and body. Just as I decide to modify and improve the information received from the discs so the soul functions. It is an emergent property of the brain and engages in independent thinking and action. The evidence for such emergence is the presence of emotion, character, aesthetic response, yearnings for God; in short what make us into persons. And these cannot be reduced to the material aspect. They identify us and could persist after death.

It is agreed that consciousness is a mystery. The brain has one hundred billion neurons. How do we do we get from the firing of neurons to consciousness? Our self-consciousness makes us aware of our difference from the environment, the need to make moral decisions and plan for the future. It dawned with *homo sapiens*. The question then becomes: did morality, spiritual awareness, desire for autonomy, arise with the human consciousness and effect our relation with the creator?

But if the brain evolved by natural selection it may be argued that capacities for aesthetic judgements and religious beliefs arose in the same mechanical way. The mind is there for survival and reproduction, and religions evolve because of the desire of practitioners to enhance their status. Morality is rooted in instinct and science will investigate how we got our values.[7] There is a theory that says we can explain everything by its origins but it is problematic, for various levels arise with new features. Complexity means that new phenomena occur which are not totally explained by lower structures. We cannot reduce biology to chemistry or psychology to biology; new categories are needed if consciousness and self-consciousness are emergent.

People see things differently. Why? It would appear that the sensory input is transformed by mind. Humanistic psychology focuses on choice and creativity, on conscious experience, on wholeness of human nature. The left side of the brain, as we mentioned, deals with language whereas the right side is more mathematical. But these areas work together. The neurons consist of nerve cells and there are three types: sensory or motor which carries information, interneurons connecting these, and synapses. The synapses are neurotransmitters or chemicals which pass through the narrow gaps.

Animals have no real syntax or grammar but use signs, visual and auditory and so on; humans have the only 'true language'. Washoe, one of the most famous of the chimps, was able to handle much communication by signs and managed about one hundred and fifty. The development of the cortex or front lobes makes the difference, with a fourfold increase in the human brain over the chimp. Memory is noted in the octopus who grabs a white ball as a reward but not a black one which has no benefit. The brain functions by the neurons working together hence the whole is as important as the parts. All function together just as an orchestra produces a sound which the individual instruments cannot. Consciousness apparently involves the whole brain.

The mind has been compared to a pattern in the brain. I am doing one thing but stop and decide to do another. This decision causes another pattern to form in the brain and physics cannot tell me why this should be. I have interrupted one flow of activity and replaced it by another. But the neuroscientist will say that having observed my brain for a long time she can see when a new pattern will form. It is a natural process but a downward causation or influence is necessary and there are occasions when we do the unexpected and unpredictable. The mind is subjective. How can science, which is objective, understand it? Science deals with observable phenomena but how can I observe my mind, how can I enter into other minds? It is not necessary to argue that the self or soul is a non-material entity embedded in the body but that it transcends it and, as Kant believed, space and time. There is no direct evidence from the senses that mind does not exist or that a spiritual world does not exist.

The computer analogy means that the software or instructions or code has evolved and is transmitted by the DNA which is modified as we respond to the environment. Thus we speak of the self which conveys the instructions, memories, behaviour patterns, rules of decision, and feelings. This software or mind runs the brain; it is a body of information, not a substance or a thing, and could run on another body after death. The mind would seem to be a global attribute of the brain and not a local product of any of its parts. Computers compute but the mind has an awareness, insight and creativity that programs the mechanical computation.

After an exhaustive survey of computers, the mathematician Roger Penrose concludes that 'mere computation cannot evoke pleasure or pain or poetry or what we feel looking at the beauty of an evening sky or the magic of sounds or hope or love or despair'.[8] Brains and organisms which they inhabit are not closed systems but in continual interaction with the world and brains are capable of modifying their structure and have the ability to make judgements. A computer is a dead storage system but the brain is living and dynamic. Penrose thinks that our consciousness is a crucial ingredient in our understanding of mathematical truth. We must 'see' the truth of a mathematical argument to be convinced of its validity. This 'seeing' is the very essence of consciousness. The operation of computers is based on the construction of algorithms which are formal procedures for computation. But the development of mathematical methods goes far beyond the formal procedures required for computation. When we 'see' the truth of a mathematical argument we reveal the non-algorithmic nature of the 'seeing' process itself.[9]

Today there is a reaction to the militant reductionism which started in the nineteenth century. It taught that the mind was to the brain as a whistle was to a train, and following this line of thinking E.O. Wilson claimed that evolutionary biology and neurobiology would do away with psychology and sociology. But his reduction is challenged by Steve Rose, who points out that reducing a page of writing to individual letters and the chemical constituents does not give us the meaning which results from combining words, sentences and paragraphs.[10]

It would appear then that the mind or consciousness emerges from the physio-chemical aspects and is not something added to them. Mind means a higher level of being and the concept of a person unites everything. There is an integration which transcends what we call the parts of the organisation, just as in Gilbert Ryle's illustration the university integrates the colleges but transcends them. A person of course does not develop in isolation from others, hence there is a need to take this into account.

Body and mind complement one another. I can account for my actions in terms of the physical processes in my brain and body or I can talk about intention and purpose. These are complementary descriptions of my behaviour not contradictory, for they concern the one person and both are required if a full account is given of an action. Just as music and its meaning express themselves through sound waves, the mind expresses itself through the body. I cannot say why music moves me by taking the piano to pieces; similarly I cannot find mind by reducing it down to the various nerve cells and neurones of the brain.

Process theology

At least we can say that the mind has an influence over the working of the brain and we might see an analogy between

God as Mind and the world. Just as our mind influences the brain, God influences the world. Process theology contends for this persuasive influence but goes further and sees the world influencing God. It contends that an absolute God unaffected by events in history is incompatible with a dynamic, progressive, suffering and relational world. We need to understand God as the leader of a community and the world as an entire web of dynamic relationships. An organism has different levels: atom, molecule, cell, organ, organism, with influence between the levels and all contributing to the whole. When we extend this to society we see a unity and interaction, with each individual contributing creativity and spontaneity. Internal relations are as important as external ones.

Thus the universe is like an enormous reverberating chamber in which any whisper echoes and re-echoes throughout the whole. A.N. Whitehead examines how a new event arises from a preceding cause but each entity handles it differently and makes a contribution. Room is left for the novel and creative. There is a creative selection of possibilities in the context of goals and aims and this is final causation. It is different from physical causation, being more like influence: teacher on pupil or the lover on the loved. God moves the human will from within, not overruling but stimulating its powers so that an act is free. His influence is the final causality: the attraction of the good. The cosmos is a unity but the mechanist separates it into parts, omits the impact of the whole, and misses the pattern. We are not some kind of substance with qualities, for no philosophy has been able to define this concept satisfactorily. In Whitehead's philosophy, being becomes becoming, and substance process, and we are persons in relationships.

God is the great persuader, companion, sufferer and leader. He is the ground of order and novelty and contains within himself the order of possibilities and potential forms of relationship. He brings pattern and order out of the process and in it he is the timeless Being, the impersonal structure of the world, passive and unchanging. But he is evolving in relation to us, temporal, fulfilling himself, affected by what happens, guiding and luring us to himself. He is the principle of actuality but influenced by the world and in this respect he is temporal and suffers with it. Causality as a physical force is replaced by personal relations on the model of human relationships. Thus there is the dual aspect of God, one personal, the other impersonal.

Whitehead rejects an initial act of creation out of nothing. He stresses continuing creation, with God as the initiator of all events; he remains eternal in his character and purpose but evolves in interaction with the world. Both Whitehead and Charles Hartshorne try to avoid the criticism that they are presenting a limited God struggling in the process. Hartshorne contends that the deity possesses both relative and absolute characteristics. Being is the abstract fixed aspect of becoming but the stress is laid on becoming. God is prior in status, being the primordial ground of everything but he is our fellow sufferer. Hartshorne embraces panentheism but does not think of God as a Person, which disagrees with John Cobb, who calls God the pre-eminent person in a community of interacting beings.[11]

We agree that God is the great persuader, companion, sufferer, leader and pre-eminent person and unchangeable in his character but there is little mention in all this of God as redeemer or liberator. The priority of status is preserved but looks like first among equals. How knowable is the primordial and impersonal aspect and does revelation really flow from it? There is a vagueness here. Process theology is in reaction to the monarchical model of God and goes too far in its modification of God's sovereignty. It requires a stronger doctrine of divine initiative and moral judgement. Critics of process thought insist that the picture of God is too weak, portraying him as a kind of 'cosmic sponge'. While that may be going too far, the picture of a God emerging in the process appears too limited and the stress on immanence and the social results in a kind of community leader. But process theology has rightly pointed to the dual model of God for it appears in the religious traditions. We find it in the Hindu philosopher Sankara, but it is not satisfactory there nor does it fulfil the criteria of simplicity since many levels can be seen in Brahman. It has at the highest level a God who is unknowable, which Ramanuja denied. Ramanuja accepted a duality but neither he nor Sankara were able to resolve it. It is one thing to say that God cannot be fully known because of his majesty and glory; it is another that his personal revelation does not correspond to what he is like. The visions of God in the religious traditions such as that of Ezekiel or Isaiah or in the *Gita* include both the majesty and personal nature of God. There are hidden aspects but these do not detract from what is being revealed.

Miracles

Can the action of God be seen in miracles, which was the traditional view? There are various definitions of miracles but David Hume regarded a miracle as a violation or transgression of the law of nature. Since then philosophers like R.F. Holland have seen them as coincidences. A child escapes death because a series of explicable physical events cause a train-driver to hit the brakes on a train bearing down on the child. It would be regarded as miraculous from the religious point of view but merely a coincidence from the natural. Hence he departs from Hume's view of divine interference with the natural order. In any case if God is always present as we hold, then how can we think of his intervening in this way?

Hume's view has been much discussed. As violations of the laws of nature miracles are impossible on the basis of experience, hence they must be rejected. But is our experience wide enough and do our views about laws of nature not change as we know more about them? Things behave in regular ways and when an exception arises to the regularity we are likely to dismiss it. But at the level of quantum physics the regularity is not observable in the random motions of particles. That is at the micro-level, but do regularity and predictability not hold at the macro? They do in general, but there are exceptions, as in chaos theory which, as we have seen, Polkinghorne tries to relate to the possible action of God. But even if we grant this, how can we trust those who report miracles? Hume wanted witnesses who are reliable and of sufficient numbers, who are not expecting wonders to happen, and are sufficiently educated not to be deluded. We will see how these criteria can be applied to the gospel miracles but it would seem that Hume started with the assumption that they were impossible, which is not being objective.

The scriptures and traditions of the various religions are full of them but the naturalistic explanation is that mistakes have occurred in reading the texts or that they were part of the culture of the time and so on. It is correct that some can be explained in this way but others cannot. Can they be treated as signs? It is done in the Fourth Gospel where the author is interested in the meaning of what has occurred. But faith cannot be based on miracles and Jesus at times refused to do them. He was aware that false prophets could work miracles in order to deceive, hence he does not use them as a proof of his divine mission (Mk 13:22ff, 8:11–13; Lk. 4:23). They are not done to enhance his prestige as a miracle worker or to meet his own need but to meet the needs of others.

Today we accept that natural laws are based on probabilities and there are reports that miracles have occurred, for example, at Lourdes and some have been confirmed by scientists. These are healing miracles but some contend that they are due to the mind bringing about changes in our bodies. Maurice Wiles is not happy about miracles, arguing that the divine action is in relation to the world as a whole rather than to particular occurrences within it, hence there will be no individual divine acts. How can we say that God acts at one time to save someone and not at another? But would we not expect God to act occasionally, otherwise we have a form of deism? Does he not act when we pray? How can we talk about God existing and not acting?

Can God act through the laws of nature just as he acts through human beings? Some event which saves one from disaster is interpreted as a miracle. If we say that God cannot temporarily suspend or transcend his own laws then we have made him a prisoner of them. Miracles are unique for while they may be regarded as 'breaking' a law of nature they do not change it. A law is established on the basis of repetition but miracles cannot be repeated. Some of them are more important than others, such as the resurrection of Christ, and we will consider arguments for and against it in this chapter and the next.

Perhaps we might think of God adapting his laws on occasion for a particular purpose. We know that it is a law of nature that iron will neither float in water nor air, yet we have ships floating on every ocean and aircraft flying. The law has been adapted. Theologians are always worried about the God of the gaps, that is, seeking to see God's action in something which has not been explained by science. Such a procedure has been embarrassing for in due course science has found the cause. If it was able to discover all the causes of the miracles of Jesus would this reduce his status? Unlikely, for if he did them by natural causes it would put him far in advance of his time.

Some miracles have been understood as later additions due to the desire to elevate the Founder of a religion, but others insist that God can suspend his laws and, in any case, quantum physics has shown that statistical laws operate: there is no rigid form of cause and effect and we can only predict probabilities. Chaos theory is an example, for the slightest change in initial conditions of a system can ensure that predictions are faulty, such as the instability of weather patterns. In a more open-ended world, science has difficulty in predicting what can or cannot happen so the theologians operate with a variety of interpretations. Perhaps it is better to think of laws being adapted rather than suspended.

The problem is that we are used to the normal hence anything extraordinary needs to be very well attested. There is the historical question: did miracles happen; and the scientific: could they happen? The historian taking into account the context may say the record is of a primitive people and such eye-witnesses are unreliable. The scientist may conclude that what appeared miraculous to them will not convince us with our knowledge of the laws of nature. But the laws of nature describe, they do not prescribe what will happen. They are based on regularities which we accept because we have observed that nature normally acts in this way. But we cannot observe all of nature, hence we do not know that it will always behave in the same way. We cannot with certainty believe that the future will always resemble the past. We are in the area of probability.

With regard to the context being primitive, if the people concerned were used to miracles, why was Paul laughed at when he spoke of the resurrection (Acts 17:32) and why did the miracles of Jesus excite such astonishment and awe? People at that time and place may have been less primitive than we think and large numbers of witnesses testified to these extraordinary events. Paul cited five hundred witnesses of the resurrection of Jesus (1 Cor. 15:6). Did they all collude with one another? When they preached it they were ridiculed to such an extent that some were afraid to report the resurrection. In Mark we read that the women having met with the risen Christ (16:1–8) 'said nothing to anybody for they were afraid'. The additional ending to the Gospel shows incredulity and dullness and lack of belief, so Jesus appears and reproaches them for not believing the witnesses (16:14). It is quite remarkable that if the writers of the Gospels had wanted to prove such an event that they should mention women, who were held in low esteem in their society and were considered unreliable as witnesses. There were doubters but Paul shows how incredible their position was in 1 Cor. 15:12–19. Miracles required prior belief in a God who acts, and in the uniqueness of Jesus and in the power of the Spirit. R. Bultmann however insisted that the resurrection was something which happened to the disciples, the 'rise of faith in the risen Lord', that is, a subjective experience. But the records indicate that it was something which happened to Jesus, not an interior event experienced by the disciples.

The miracles of the Gospels must be investigated individually to see if they resulted from natural causes. If we consider the miracle of the loaves and fishes it is difficult to accept that Jesus literally multiplied them since he refused to do this according to the record of the temptation (Mt 4:3–4). The crowd will always follow the man who gives them bread but for the material benefit not any spiritual need. Hence various interpretations have been put forward: it may have been a sacramental meal with a morsel like the sacrament or it may be that the crowd had brought food and the miracle was that they shared it. But these naturalistic explanations do not take account of the reaction of the crowd. Something remarkable seemed to have happened that encouraged them to want to make him a king (John 6:12–15).

With regard to the raising of the dead it can be explained naturally since bodies decomposed so quickly in that climate that they could be hastily buried while in a coma. But Lazarus was in the tomb a number of days. It is strange that the miracle is not recorded in any of the other gospels. How could they have failed to notice it? Hence some have considered the account an allegory written as a result of Jesus saying that he was the resurrection and the life. Some commentators think that we should use Mark as a test of what the other gospels say, but he too has miracles. Another theory is that a miracle is God using natural resources not suspending them, as in the crossing of the Red Sea, where it is stated that the Lord drove the sea back all night with a strong east wind, thus turning the sea bed into dry land (Ex. 14:21).

One of the most debatable miracles which is accepted by the Church is the virgin birth. Critics say that *almah* means young woman, and if Isaiah had wanted to stress a virgin he would have used the word *betulah*. The Greek translation uses *parthenos* which can mean virgin but its basic translation is girl. Both Matthew and Luke believed in a virgin birth but Luke does not refer to the Isaiah prophecy and uses *parthenos*. Perhaps there was a misunderstanding in the Greek version (Septuagint) of the Hebrew scriptures but the linguist *qua* linguist cannot prove it happened.[12]

Keith Ward says that Jesus is the son of God not in the physical sense but metaphorically indicating a unique relationship with God. He believes in the virgin birth and notes that it is possible for women to give birth to females by pathogenesis, that is without insemination by a male. But women do not have the Y chromosome necessary to have a male child so it is not possible within the limits of laws of nature hence these laws must have been suspended. Since the life and death and resurrection were unique and miraculous it follows that his birth would be of the same quality, therefore Ward is inclined to believe it. But if the literal truth is denied, as Ward notes, we can still retain its symbolic value since the miracle is not essential to the truth that Jesus was the Messiah of Israel. He thinks that Jesus in his natural and healing miracles transcended the normal powers of nature, but all the miracles have a symbolic nature and depend on accepting that God was incarnate in him.[13]

Polkinghorne does not dismiss the virgin birth and opposes Peacocke. The latter notes that Mary could only supply an X chromosome so Y must have been miraculously provided; but if it happened it would affect his true humanity. Polkinghorne however is not convinced and contends that Mary's conception was an act of divine/human co-operation, that is, the power of the Spirit and human obedience.

Another debatable miracle is the walking on the sea. Is it possible that it was a misinterpretation of the event and that he was walking on high ground and came down to the shore to help? In Matthew 14:25 it is *epi ten thalassan,* which can mean over the sea or towards the sea, and in verse 26 it is *epi tes thalasses* which can mean on the sea. The phrase is used in John 21.1, meaning by the sea shore. The same word *peripatein* (walk about) is used in these versions. But what are we going to do if these theories are accepted with Matthew's account of Peter attempting to do it and sinking? And there is the awe and fear: 'Truly you are the son of God' (Mt. 14:22–33).

In an attempt to find non-physical causes for miracles some have investigated psychic phenomena. They contend that while a miracle is a law-abiding event it releases energies which are normal on a plane of being higher than any with which we are familiar. A distinction is made between *blepo* and *theoreo,* which come through the optic nerves, and *horao,* which is mental insight or spiritual vision. In John 16:16, 'A little while and ye shall not behold (*theoreo*) me and again a little while and ye shall see (*horao*) me because I go to my Father'. One signifies the physical presence of Jesus and the other his spiritual or resurrection presence.[14] It means that the resurrection was a spiritual event, not physical, but the record insists that he did partake of food (Lk. 24:30) and there was the physical appearance to Thomas (John 20:27).

The healing miracles often required dealing with the demon-possessed. At that time belief in demons was widespread and resembles what is encountered in villages in India today. Some argue that Jesus believed in them because he did not have our medical knowledge or since the sufferer believed, Jesus assumed their beliefs in order to cure them. One of our current problems is that while we understand psychosomatic illness in a different way there still remains illness without any bodily cause and it remains a mystery.

The Gospel records

What appears clear from the Gospels is that we have a unique person who does extraordinary deeds; but how he did them is difficult to discover. In the case of such a person with a unique relation to God we would expect the extraordinary and that is what the New Testament said happened. But the historian is anxious to know not only where the miracle stories originated but the whole story of Jesus. Form criticism tries to get at the material in the Gospels which stem from Jesus and not the thinking of the early church. Mark is the earliest passion narrative but the rest is composed of units aranged by subject matter more than by historical or chronological concerns. The units are pronouncements, stories, miracles, parables and aphorisms adapted to needs of the post-Easter community and circulated in oral tradition. Both Matthew and Luke used Mark plus oral traditions. All sources were shaped by redaction.

Three stages can be noted:

i Authentic words and memories of Jesus himself.

ii Materials shaped and transmitted in oral tradition

iii The evangelists' redaction. The fourth gospel does contain some stage one and stage two material but generally has stage three material.

The criteria for stage one is multiple attestation, that is, more than one source of material which is distinctive to Jesus since there is no parallel in Judaism or the post-Easter community. It confirms rather than excludes the principle of dissimilarity and coherence with other Jesus traditions. In the case of sayings there are signs that they originated in Aramaic for this was his normal language. But there was a difference between the teaching of Jesus and the rabbis with his stress on love of neighbour and the enemy. His arrival meant that the age of Moses was coming to an end, for the kingdom had come and God's original intent in creation was being realised. The message is mercy and forgiveness (Mk 1:15), continuing creation (Mt 6:26, 30–2), and God acting in history. He addresses God as Abba for he is his Father in a personal sense. The new appears in the miracles of Jesus, for he raised the dead. The material belongs to stage two but the resuscitations may go back to stage one. The stories may rest upon a general memory that Jesus did perform such deeds and it is noteworthy that the feeding of the multitude has multiple attestation being in three independent traditions.

Jesus dared to speak and act for God in the phrase: 'I say unto you (Mt 5:21–48), he forgave sins which only God could do (Mk 2:5–12), and acted with authority. But he did not claim to be the Son of God in any metaphysical sense. These passages belong to stage two or three but his response of full obedience indicated his unique sense of sonship. It was not metaphysical sonship but showed an historical call and obedience. If Mark 10:45 belongs to stage one then Jesus is modelling himself on the suffering servant (Isa. 53). There is no explicit Christology but what God did through him in his earthly ministry provided the materials for the christological evaluation of Jesus after the Easter event.[15]

There are passages which resemble the fourth gospel particularly, such as Mt. 11.25–7, which has been called the Johannine thunderbolt. It indicates the very personal relation of the son with the Father. Then, on the mount of transfiguration, he meets with Moses the greatest lawgiver and Elijah the chief of the prophets and they acknowledge what Jesus is doing and he is the one for whom they have been waiting. He is lord of the sabbath and there was nothing more sacred to the Jews than that day. Knowing this the Romans chose it and easily defeated the Jews in battle (1 Macc. 2:31–8)

But it is in the Fourth Gospel that sonship acquires a metaphysical meaning[16] with Jesus being called God (1: 8f, cf. 1 John 5:20). Apparently a high christology developed within a short period. The writer of the gospel makes great claims for Jesus and critics have written these off but at times they appear in the Synoptics. Luke repeats the passage from Matthew which looks as if it should be in John: (Lk. 10:21–4): 'All things have been handed over to me by my Father. No one knows who the Son is except the Father; and no one knows who the Father is except the Son and he to whom the Son wishes to reveal Him (cf. John 1:14, 10:30, 14:9). The charge which brought Jesus to his death was blasphemy and the accusation that he was a political agitator. The Jews put forward both accusations, since if it had been blasphemy only, Pilate would have told them to deal with Jesus themselves; but they were unable to impose the death penalty.

Incarnation

The church saw God acting in Jesus and eventually arrived at the doctrine of the incarnation. In some religions the concept happens repeatedly and some see it as an ancient mythological form of thought while others as the entering of God into human life. But it would appear to be the distinctive core of Christianity.[17] If it did occur then the experiment was renewed in him. Something new entered into it with the coming of Jesus. The parable of the old and new wine is appropriate here. The religious leaders of the day were used to the old teaching and had minds closed to the new. They argued that the old is better, as wine usually is, but forgot that the new would mature and might be best of all.

God acts through Christ, other religious leaders and communities, and the Spirit who guides and directs. Mankind differs from the animals in having a spirit-given faculty to recognise goodness and truth. But what is not used can be lost. Prayer is also regarded as the way to influence the action of God, for the divine plan is an open one. God does not overwhelm us but persuades, often through others, to do his will.

In conclusion it would seem that faith is essential if God is to act. The record shows that Jesus could not do miracles where people did not believe in him. The parable of the leaven (Mt. 13:33) seems to indicate that the action of God is unseen. Just as we cannot see the leaven working in the dough the purposes of God are invisible but realised in their effects. But other commentators hold that we do see the leaven having an effect on the dough since it seethes, bubbles, heaves, and in similar manner the kingdom can come with a great disturbance (Acts 17:6). Perhaps the two views, depending on the circumstances, can be held together.[18]

In the next chapter we consider if there is life after death.

8
The End of the Experiment

We consider in this chapter the picture painted by science of how the world will end and the various views of what form life after death might take if there is one. What is the religious picture and how do we understand the apocalyptic view which emerges?

Stars and the sun are seething nuclear explosions and already the stars are beginning not to form as the universe runs out of hydrogen fuel. The earth will die before the universe as the sun gets warmer and our descendants will need to escape to some other planet. But it may be that the ashes of our sun and the earth will form the raw materials for new solar systems. Such recycling could bring the elements necessary to sustain life: carbon, iron, oxygen, and so on. But there may be life elsewhere in the cosmos for there are many stars whose elements generate it. When the universe is one hundred times older than it is now the recycled fuel will run out and black holes will take over. With the explosion of many stars there will be a vast number of them and soon after this our universe will die. What began with the Big Bang will, as T.S. Eliot said, end with a whimper.[1]

The sun is an amazing ball of fiery gas made of hydrogen and helium, very light atoms, and it orginated about 4,600 million years ago. The nuclei of hydrogen and helium crash into each other and give off energy and heat. The sun is further testimony to the regularity of the universe since it manages to feed the hydrogen fuel in its fire at just the right rate. When this fails to happen in about 5,000 million years all life here will be extinguished.

The earth goes round the sun in an orbit with the huge force of gravity preventing it flying off into space. The sun does not stay still for it belongs to a galaxy which is rotating and it takes about two hundred million years to make a complete turn. The earth itself is moving at four hundred kilometres per second and goes round the sun in a year. It was formed at the same time as the sun and the other nine planets and is not solid but just a thin floating crust which churns about. Cracks form, resulting in 'plates' as we noted, which move and cause earthquakes. A volcano is a weak spot in the earth's crust and pressure builds up on the molten rock causing it to shoot out.

Gravity is a key factor in the end of the universe, slowing matter down and eventually leading to the Big Crunch. If there is little matter or an anti-gravity force then it will expand forever. This may be correct but even if it continues expanding, stars will have disappeared and the black holes will evaporate leaving a universe in darkness.[2]

Replicas

Some advocate escaping to other planets and learning to live under different conditions. Freeman Dyson contends that life is a matter of organisation rather than flesh and blood, so consciousness, which results from such organisation could be reproduced in other forms. It will be necessary if we are to survive the new conditions, particularly if the universe is going to die by slow freezing rather than heat. The basic point, as we mentioned, is that the mind is a software program within the hardware of the brain and could be transferred to other hardware. The idea is also embraced by Frank Tipler who engages in much speculation. He thinks that we can escape to a supra-physical dimension of immortality using the mechanism of computer simulation.

We are information processors and could be reduplicated or replicated for the soul is like a complex computer program. The risen Jesus was a simulated person appearing and disappearing in different localities and he was the first to do this. Tipler is not a Christian but thinks that religion is now part of science. Unfortunately there is no redemptive significance in his scenario and it is clear that we cannot be reduced to information. We are more than that and can a replica equate with what we are? It opposes biblical eschatology, which means the recreation of humanity.[3]

Cloning has contributed to the discussion of life after death. It might be one way of conferring a kind of serial immortality on those who die and is dependent on the view that genetic identity is equivalent to personal identity. But a clone created with resurrection in mind would be of value for its conformity with his progenitor, that is, identity in personality, behaviour, talents and so on. We are more than the sum of our genes, and nurture as well as nature makes us what we are. Genes predispose but do not determine us. Cloning is not copying and in any case it is original paintings which are valued so highly, not copies.[4]

Near-death experiences

Some people think that these may confirm the hope of immortality. Research is being done into whether the mind can have experiences when the brain has ceased functioning. The usual report is that a tunnel and light have been seen which gives a feeling of peace and love. Critics question these experiences and argue that these people are not dead, but doctors report that the brain has stopped functioning. In the mid-1970s Raymond Moody recorded experiences of fifty people and the experiment was continued by the psychiatrist Brian Grayson. In such a condition thinking accelerates and there is a feeling of popping out of the body. It is an extra-sensory experience.

Work is continuing in Southampton on cardiac arrest where there is no heart beat and no brain function. Dr Peter Fenwick says that all experience should stop when it occurs but Susan Blackmore, who is sceptical of the research, thinks that the big question is, when do supposed post-cardiac arrest experiences really occur? She thinks, before we go out of consciousness and when we come back into it. But this is speculation. The figures produced show that of 63 patients only four had near-death experiences but on a wider survey of 300 patients, 41 said they had them. One particular example is striking: Pam Reynolds. Her brain was not functioning during an operation yet she knew that she was looking down at the body. She heard voices and saw the doctors and the instruments they were using during her brain operation. Later, when conscious, she told them and they confirmed what she had seen about their activity. How did she see and hear when her senses were cut off? Or could it have been that she was familiar with the work of doctors during such an operation? She also recalled seeing relatives and a great light and was told that it symbolised God.[5]

Roger Penrose is investigating the depth of our brain cells and is looking for a pattern deep within brain consciousness. If such a pattern exists it has a similarity with what Polkinghorne is talking about. The question is, can information leak out after the brain stops functioning?

Penrose is also interested in quantum effects in the brain and his research shows the difficulties this poses for understanding it. Quantum particles contain information which makes them behave differently from the normal. When we are not measuring an electron it may do things in a clockwise direction and another in an anti-clockwise: up or down. It opens up the possibility of science fiction becoming reality in the teleportation of particles. Electrons in an entangled pair are connected even when separated by a distance so that a quantum state of one electron can be moved from one particle to another. It sounds strange, but this is the sub-atomic world.

Philosophers engage in a criticism of Platonic and Cartesian dualism of body and mind or soul. This idea of the immortality of the soul is still held by some philosophers and makes the argument for life after death easier. But a life in non-bodily terms does not agree with what we have noted about the Judaeo/Christian belief in a psychosomatic unity. Immanuel Kant put forward the moral argument for life after death. If God exists the highest good will finally be realised, not here but hereafter. His justice requires it but there is also an objective moral law. Today when so much is relative such an objective moral law has been questioned. Morality, it is said, is a matter of feeling and emerged to hold societies together. It continues to be a matter of debate, as does the question whether life after death is desirable.[6] The believer has no doubt about it, for he believes that 'eye has not seen nor ear heard what God has prepared for them that love him' (1 Cor. 2:9). But a new body is necessary when God will wipe away all tears from those who have grievously suffered in the present one (Rev. 21:4).

The religions

Turning to the religions we see that there are two concepts of time: linear and cyclic. The linear view sees a beginning and an end whereas the Indian belief is of the world being destroyed at the end of periods and then again recreated. This cyclic view is also found in Greek Stoic philosophy.

Immortality has been hoped for by mankind from the Neanderthals 100,000 years ago who placed necessities in graves, which no other species did. The religions seek to confirm this hope. In Judaism the goal is to obey the Torah, the commandments of God, but also to look forward to the perfect state when the kingdom of God is established with the coming of the Messiah. It is not only national but universal liberation. The new aeon will mean the judgement of the evildoers in the resurrection of the dead and the enjoyment of a closer communion with God for the obedient. Hence the Jews have an other-worldly goal: the coming of the Messiah who is human from the house of David, the return to the promised land, the victory of God over their enemies, and the judgment of the world. Everyone has a soul but it is not the Greek idea which disparages the body. Both soul and body are closely united so there will be a resurrection of the body as in the Christian belief.

Judaism in the post-exilic period was influenced by the Platonic idea of the immortality of the soul (Wisd. Sol. 2: 23–34). But it was also thought that in the messianic age there will be a bodily resurrection.[7] However, Reform Jewry does not envisage the coming of a personal Messiah and has deleted reference to the resurrection of the body from their prayerbook and inserted 'life eternal'. They believe in the permanence of the soul. Cohn Sherbok, who mentions it, seems to have set aside belief in the afterlife when victims of the Holocaust will be rewarded. The reason is the mystery of God's dealings with us, since religious traditions are in the end nothing more than lenses through which we perceive reality. God lies beyond our comprehension and the puzzle of his providence during the Holocaust is an unfathomable mystery, so we should cease to agonise about it.[8]

It is true that a few verses of the scripture oppose life after death but it came to be believed probably around 165 BC that there was some sort of existence in Sheol (place of the shades or weak ones: *rephaim*). A fuller hope is in Daniel (12:2). which was written in the Maccabean period and has been interpreted as meaning the resurrection of the individual righteous man. In the Psalms the view is taken that death cannot interrupt communion with God (Ps. 73:25–6) but the life after death is still regarded as the weakest form. The Hebrew idea was that the reward for doing good came to man in this life so tension was set up when it did not happen. Reflection strengthened the belief in an afterlife, hence the idea of hell and a rather mundane heaven is found in the pseudonymous apocalypses (Enoch 21; 22:10f, 11; Esdras 7:36). But in the Wisdom of Solomon there is a sure hope of immortality for those tested by God and found worthy to be his (3:1–5). The more developed doctrine of the resurrection of the body was denied by the Sadducees in the time of Jesus but held by the Pharisees probably on the basis that justice must be seen to be done elsewhere. The Jews do believe that God is leading them into a new future: the new exodus, the new Jerusalem, the new covenant, the new creation and consummation.

The Muslim looks for a paradise with rivers of milk and wine and those who reach it will wear robes of silk, recline upon soft couches, and be attended by boys graced with eternal youth. Such a vision inspired them to go into battle, for death meant immediate entrance to it. In the final assessment, all who die recover their bodies and stand before the Judge and to each is passed a book. If it is placed in the right hand it is the passport to paradise, if in the left it means hell, which is described in vivid imagery. But if Muhammad prays for some of them they can be redeemed. The Shi'ah look forward to the revealing of the 'hidden Imam' but the Sunni stress the day of judgement, not any apocalyptic intervention in history. Buddhism too speaks of an ultimate nirvana and in the Mahayana tradition there is the blessed land.

Hindus believe in life after death, with the *Veda* affirming that the ancestors enjoy a blissful existence, but the reincarnation belief acknowledged as a universal fact in the *Upanishads* means that there is need to escape the cycle of life and death and merge the soul with Brahman, or the transformation of the individual by belief in a personal god such as Vishnu. There are hells for the ungodly and some who have not had the proper rituals performed for them cannot find rest until they are done.[9]

Hinduism centres its belief on the concept of self which 'is never born nor dies ... nor will it ever come to be again: unborn, eternal, everlasting is it. ...' (*Gita* 2.20). The eventual escape from the self or soul from the body is the hope. The *Bhagavadgita*, having been written after the emergence of Buddhism, does reflect its teachings at times when it says that the escape necessitates the sweeping away of the ego. It is necessary if the true self is to return to its spiritual home, 'Nirvana which is Brahman too' (2.72). But the *Gita* does not deny the existence of the self, for it is part of God and requires detachment from the world and intense concentration (*yoga*). Otherwise it is deluded into thinking it belongs to the body. The problem arises because the body material of which we are composed can blind the self to its true nature.[10]

Buddhism believes that we are composed of five aggregates: matter, feelings, perception, mental formations, consciousness. What is strange is that consciousness, or the mind, is not given the central role which appears in other religions but is just another element. Yet the mind in Buddhism has a leading role in meditation, so should it not be more central?

The goal is *nirvana* which is peace and insight in this life and, after death, freedom from rebirth. But *nirvana* is not easy to understand and there are both positive and negative views. How can the individual survive after death if she is simply a succession of mental and physical states? The Buddha was ambiguous about it and did not like such questions! The root meaning of *nirvana* is 'to blow': blowing out or the extinction of self. There is an affinity with Christianity with its denial of the self or death to it.

But the self or soul is illusory in Buddhism so what is reborn is the web of *karma* in its causal connections. Just as one flame is passed to another or a billiard ball striking another stops dead while the other ball moves on, strikes another and itself stops dead, then a third continues the process. But the Buddha moved beyond *karma* as action to intention. It was the intention of a deed which was important to him and this, of course, is crucial in determining the punishment for any crime.

Another explanation is to see that the aggregates or *Skandhas* relate to each other according to patterns which are reproduced from one moment to the next and provide the continuity of the person. Death interrupts but the *Skandhas* form new patterns which are not identical to the old but connected with them. Thus the new being that is born is not identical with the old but not completely different. We are reminded of what we said about replicas, but *karma* ensures the causal sequence.

The Buddha denies both the permanence of the soul and any immortality of the gods or God. The one permanent thing is *nirvana,* which is the extinction of the fires of greed, hatred, and the delusion that we have a permanent self. Once attachment ceases, *nirvana* is attained. It can be attained in this present life as is shown by the *Arhats,* worthy persons, but there is an ultimate *nirvana,* for the Buddha spoke of 'another shore', 'a deathless realm, unborn, uncreated ...'. While all other things are caused or conditioned, *nirvana* is unconditioned. Philosophers continue to debate whether it is positive or negative but perhaps the best view is that *nirvana* is not a place but a state and since it belongs to the area of the absolute it cannot be described, being without properties.[11]

Biblical eschatology

Jesus predicted the destruction of the Temple, which occurred in 70 CE, and instituted a new covenant for those who would follow him, mentioned false messiahs and natural disasters, and promised that he would come eventually as the Son of man (Dan. 7:13, 14). The letters to the Thessalonians deal with the second coming, with Paul insisting on events that must occur first. He refers to apostasy and the man of lawlessness and the force that is presently restraining it. Perhaps he meant Rome, which was responsible for order in the world, or that God was using the preaching of the gospel to hold back the full impact of evil.

The force of evil is vividly depicted in the Book of Revelation and the battle with the forces of God but it appears also in the ministry of Jesus and a striking depiction is the parable of the tares in the wheat (Mt 13:24–30, 36–43). The tares are mixed with the good seed but the reapers must wait until the harvest to separate them. God is the reaper and judge at the end and the new creation will redress the balance of the old. Debate has ensued about statements made by Jesus with respect to his future coming. According to Matthew it was to occur within the lifetime of those who listened to him but perhaps the real meaning is in Mark 9:1 where we read, 'Truly, I say to you, there are some standing here who will not taste death before they see the kingdom of God come with power.' Such power was demonstrated in the descent of the Spirit at Pentecost and in the remarkable advance of the early church. In any case, as Jesus himself said, no one but God knows when the end will be.

Biblical eschatology envisages the recreation of all things as the experiment comes to a close. But the hope is not tied to the end since it has already come with Jesus in his preaching of the kingdom of God and resurrection and his presence in the Church by the Holy Spirit. His resurrection, which we must consider, is the guarantee of ours. Beginning and end are orientated to what Christ has done (Rev. 1:17–18). Despite objections to the use of power at the end it is necessary to draw the experiment to a successful conclusion for there will always be those who continue to be disobedient to God. This would appear to be the message despite the difficulties with regard to the interpretation of the symbolism of the book of Revelation. The Spirit who raised Christ from the dead will raise us (Rom. 8:11) and, there will be a new creation of the cosmos and humanity (Rom. 8.19–23). The cosmic Christ (Col. 1:16–17, Rev. 21:1–4) will act as judge but there is also self-judgement (John 3:18–21).

Some scholars, however, doubt any kind of literal second coming and argue that the sayings about the Son of man coming are not authentic to him or that he intended them as symbols of God's triumph. But in whatever way they are interpreted it is possible to affirm the basic triumph of God over evil without commitment to any views about the nature of the second coming or when it will be.[12] The apocalypse is clothed in prophetic imagery but the idea is the transformation of all things. God is a God of hope (Rom. 15:13). But after the Enlightenment such a hope was regarded as the product of ignorance and fear, and faith was placed in the progress of humanity. Such a dream was shattered in the world wars of the twentieth century and the threat of nuclear war. Our postmodern world has little confidence in reason and even in science.[13]

The righteous will enjoy heaven, which does not necessarily mean a literal place since we read that the Christian sits with Christ in the heavenly places (Eph. 2:6). Hell is a translation of the Hebrew *sheol* for souls after death (Gn. 37:35, Acts 2.27), a place of punishment. But hades can be neutral where all the dead are kept (Mt. 16:18, Acts 2:27, Rev. 1:18, 1 Cor. 15:55). Will God ever give up trying to persuade the disobedient? John Hick thinks not and universalism might be a possibility. Jesus did teach that sinners would be punished but not with eternal torment. *Aiwnios* is ambiguous and Jewish apocalyptic themes have entered into the gospels. Suffering must be redemptive as is seen in the doctrine of purgatory. Both Polkinghorne and Hick appear to be sympathetic to it but Catholics insist that purgatory is for those who die in a state of grace. The judgement of all will be based on helping those who were poor and needy (Mt. 24:1–31), thereby showing the compassion of Christ.[14]

The present universe has properties that enable it to make itself, with death as the cost of new life. God interacts in hidden ways with the cosmos, being transcendent and immanent, but in the new creation there will be a more intimate relation with the creator. The old creation is involved in an experiment, for it is allowed to explore and realise its potentiality without divine intervention but those who have been redeemed will participate in the new creation where there will be no more evil.

The resurrection of Christ

The resurrection of Christ is the basis for the Christian hope and we need to look at it more closely. The Jews thought of resurrection in terms of glorious heavenly beings (Dan. 12:3; Wisd. 3:1,7) or the restoration of someone to his normal human state such as the widow's son by Elisha (2 Kings 4: 8–37) but the accounts of the the risen Jesus do not fall into either of these ready-made formats:

> This faces us straightaway with the following question: if the disciples were inventing the story that Jesus had come back to life or if they were simply trying to convey an inner certainty that his soul, like that of John Brown, went marching on, is there not a probability that they would have used, albeit with minor variations, a convention recognizably akin to those already current in their world of thought?[15]

The act of God in the resurrection of Christ fits into the pattern of the acts of God which we have seen in Judaism. God grants liberty to mankind and the world but his justice operates eventually when freedom is abused by evil. Thus he acts in the exodus from Egypt to liberate the Hebrews and subsequently throughout their history. When trust is placed in him he responds, and evildoers pay: death was the penalty for the Nazis who created the concentration camps and a devastating destruction of Germany. A new future has been opened to the Jews in the state of Israel and they will be punished if they do not find a way to be just to the Palestinians. Death could not be the end for Jesus; the innocent must be vindicated. His identity also needed to be revealed and the resurrection confirmed that he was the Son of God (Rom. 1:4)

It is argued that without the resurrection there would never have been a Christian church since the cross meant failure. Those who defend it ask: How could his followers have continued to tell a lie and be willing to die for it? His resurrection appearances could not have been hallucinations since they would have arisen from expectations and the record shows that they did not have them. It is impossible to think that God would not vindicate him or abandon the man who fully trusted in him, for evil frustrates the purposes of God. Yet some doubted (Mt. 28:16–20, 1 Cor. 15) and there was the problem of recognition: a gardener (John 20:14–16), a stranger (Lk. 24:13–32), a ghost (Lk. 24:36–40), a figure on the beach (John 21:4–7). There appeared to be continuity and discontinuity between the earthly and risen Christ.

Karl Barth does not think that historical evidence can support or refute the resurrection since it is an act of God and therefore historical in a unique sense. Wolfhart Pannenberg disagrees, for it is a public fact, but Barth's point is that it is not an event like Caesar crossing the Rubicon. It is *Heiligeschichte* or some kind of holy or superhistory and cannot be demonstrated by the usual historical method. Pannenberg argues that historical events can reveal God but are known fully at the end of history. We will explain this more fully later. Some scholars following Barth conclude that such an event is unique and is beyond proof or disproof but the Acts of the Apostles speaks of Jesus appearing to his disciples in space and time and convincing them by many proofs (Acts 1:3). Paul says Jesus appeared to many (1 Cor. 15:6–8). Debate among Christian scholars usually centres around the mode not the fact of the resurrection. Was it the dematerialization of a physical body and its reappearance as a spiritual one?

What the resurrection appearances show is that there was some connection between the old body and the new. The disciples recognised him by his actions which reflected how he acted in his earthly life (Lk. 24:30–1). It is interesting to note that when we see someone after a long time who has changed physically we do not at first recognise him or her. But then an action, how they walk or speak, suddenly induces recognition. Mary did not recognise Jesus at the tomb but when he spoke her name she immediately did so. In order to establish a connection between the old and the new, Paul uses a number of analogies in the First Letter to the Corinthians, Chapter 15. There is, for example, the seed and the plant: the seed sown is related to the new plant that sprouts but the new sprout has a different and genuinely new body. Our bodies disintegrate and die just like the seed but there is resurrection with a new body, hence Paul says: 'flesh and blood' cannot inherit the kingdom (1 Cor. 15:50). The explanation of the resurrection: visions, hallucinations, spirit, imagination, apparition ... are not generally in accord with the gospel record unless we are to spiritualise the appearances. But how could a spirit vision point out a shoal of fish to fishermen or light a fire or cook a meal and share it out (John 21.1–14)?

New Testament scholars have not been able to disprove the historical accuracy of the appearance narratives, according to William Alston. Susan Oakely, in reply, says that he is inclined to be too literal in thinking of a bodily resurrection, for this body was different from ours. Metaphor is to be taken into account. Swinburne stresses the importance of background evidence and contends that God may at times set aside natural law, but others believe that the resurrection brought new awareness of his presence and empowerment with no need for an objective event. Some think there may be a parallel in near-death experiences (which we have mentioned) but this seems hardly relevant since the appearances were to people who were not in that condition! What we do know is that the experience transformed the disciples and led to a belief for which they were prepared to die. It is the effects that are convincing. Visions could be possible but that does not rule out their reality.[16]

The physical nature of the resurrection is recorded in Luke 24:33–44 and John 20:14–29. S.T. Davis argues that what they saw was a material object with their eyes opened by the grace of God or the Holy Spirit.[17] Perception was enhanced not only to believers but doubters like Thomas and persecutors such as Saul of Tarsus. The new body could appear and disappear, so did not conform to natural laws. The Greek word *ophthe* can be used of spiritual seeing but also ordinary seeing of a material object (Acts 7:26) Normally *horama* and *optasia* are used for visions of God or angels (Matt. 17:9 and Acts 9:10, 16:19). Since there were five hundred witnesses according to Paul it must have been normal not subjective seeing. The eyes of the disciples were opened on the Emmaus road and they reacted with surprise and awe but there was slowness of heart to believe (Lk. 24:25). The appearance of Jesus to Paul differed from these since Luke limits them to the time between the crucifixion and some forty days later (Acts 1:3). Paul receives a heavenly vision, *optasia* (Acts 26:19).

The early church interpreted the appearances by means of ordinary vision as seen in Ignatius of Antioch, Justin Martyr and so on. But the literal, it is said, cannot be applied to the appearances since he had transcended the categories of space and time. To interpret them by metaphor and symbol opposes the record of the Gospels, which states that he did physically appear to his disciples. R.H. Fuller accepts that the report of the empty tomb has a factual core and that Jesus was restored to life but is not happy about a corporal body. In reply W.P. Alston argues that the appearances are not like the Pauline vision but corporal, for the record says: 'see my hands and feet that it is I myself' (Lk. 24:39). He criticises the criterion of dissimilarity, that is, accepting only material which is dissimilar to that of ancient Judaism and the early church. It would leave us with little knowledge of Jesus.

Fuller does not deny that there is an historical basis in the gospel accounts and that it is reasonable to accept that he did appear to the disciples. This is the main point; the form used is secondary. The failure of New Testament criticism is that it too easily accepts that the early Church added to the tradition. It doubts the theological intentions of the evangelists and believes that stories originated because of story-telling proclivity. More radical forms of such criticism go so far as to assert that the words of Jesus originated from a Christian prophet in the early Church and that the Pauline vision was typical of all resurrection appearances. The basic problem is the non-acceptance of divine interaction and the complete reliance on natural causes. R. Bultmann was prominent in this argument but Fuller accepts miraculous occurrences, though he rejects the bodily form because of some aversion to crudity. However that may be, the gospels are realistic, saying that some doubted (Mt. 28:17) and others did not recognise (Lk. 24:13ff).[18]

But Fuller apparently would agree with Barth since he writes that the resurrection is not historical in the sense of being an ordinary event because it occurs where history ends and God's end-time or kingdom begins. No one saw God raise Jesus from the dead; it is an inference from the disciples' experience and they are not to be treated as hallucinating though he thinks they may have seen visions. But he admits that the empty tomb stems from a version that is very early and the disciples verified the women's discovery (Lk. 24:12). It would seem then that the bodily resurrection is not to be easily dismissed and it is stressed by the Fourth Gospel: the wounds of the nails and hole in the side were there (John 20:24–9). The resurrection of Jesus is the first fruit of the corporate resurrection.

The most convincing arguments are the change in the disciples. How could they have proclaimed such a message with a dead body on their hands? Such a position would not have justified the joy with which they proclaimed the gospel but would have reduced them to misery. The development of the church against all odds is a remarkable feat in itself. Then there was the departure from the Jewish Sabbath and the decision that the requirement of circumcision would not be enforced upon the Gentiles. They moved from the custom of celebrating the traditional Sabbath, which for Jews was a tremendous decision, to celebration on Sunday as the day on which Christ rose from the dead. This showed the movement from the Faith in which they had been reared to a new religion which nevertheless retained its roots in Judaism. The resurrection is in the earliest tradition: (Rom 10:9, 1 Cor 6:14, 15:15; Acts 2:32, 13:34; Col 2:12, 3:1; Eph. 2:4). The appearances and empty tomb are later. Chapter 15 of First Corinthians was written five years after the crucifixion. Even P. Sanders can write: 'That Jesus' followers (and later Paul) had resurrection experiences is, in my judgment a fact. What the reality was that gave rise to the experiences I do not know.'[19]

John Polkinghorne thinks that the empty tomb is important for it signifies the transmutation of the body into the glorified one. Paul does not mention it but does record that Jesus was buried. All the gospels state that the tomb was empty. The descriptions of the appearances varied but all agree on the difficulty of recognition. The resurrection fulfils three conditions which correspond to a coherent understanding of divine action. It was not fitting that the life of Jesus should end in failure or that God should abandon the man who wholly trusted him. It was right that the hope that death is not the last word should be vindicated. Its timing is important because Jesus anticipates within history a destiny that awaits us and shows that matter will be transformed not reassembled.[20]

Life after death

The matter of our bodies is changing all the time for the atoms that make it up are not the same today as they were a few years ago. But the patterns these atoms form remain and make for continuity and that is what the soul is. When we die God will re-embody that pattern in the final act of resurrection. The soul is 'me', the information-bearing pattern, the organiser of matter. Polkinghorne thinks now that the computer analogy is somewhat crude but still of some use. It is the pattern in the brain which establishes personal identity but it is dissolved at death. What happens is that it will be held in the divine memory and ultimately re-embodied. But it might be thought that there is something nebulous about being held in the divine memory. When I think of relatives and friends who have died I do more than remember them but have the hope that they are alive in some way. The idea of being eternally remembered by God comes from process theology as put foward by Charles Hartshorne. Our life is retained in the divine memory but we do not live on after death as conscious agents. It conflicts with Hebrews 11 and other scriptural passages which insist that those who have passed on are living witnesses watching and encouraging us in the race of life.

But Polkinghorne is contending that it will only be for a period, perhaps in an intermediate state until the final general resurrection, when we will become alive again. All will depend on the faithfulness of God. But does entry into the future state not take place at death? (Lk. 16:19ff) There is some doubt here since it is also taught that the dead sleep in the tomb until the last trumpet. But the penitent thief is assured by Jesus that he will be with him immediately in Paradise (Lk. 23:43). If it is only a waiting room does it not affect the words of Jesus, which seemed to promise more than this?

Keith Ward thinks that some will never want to turn from evil so the possibility of hell exists though God desires all to be saved (1 Tim. 2:4) and offers eternal life to those who have not heard of Jesus. Since God is unlimited love, purgatory is a possibility but needs extending to those who find themselves in hell so that repentance is available. *Sheol* is a shadowy existence translated wrongly as hell in the Apostles' Creed. *Sheol* and paradise come before the final resurrection of the dead, with Jesus descending into hell to preach to those condemned at the flood. They are challenged to repent and believe the gospel (1 Pt. 3:19–20).[21] But even if this is acceptable it was to those who had not heard of Christ. Is Ward too optimistic about extending purgatory? Will we not be judged by deeds done in the body and changed as we enter into the new state (1 Cor. 15:51–2)?

The debate about eschatology

C.H. Dodd proposed realised eschatology in the ministry of Jesus. Jeremias disagreed and suggested inaugurated eschatology: Jesus had initiated a process destined to work itself out to the last day. If Jeremias is right then Weiss, Schweitzer, Dodd, Bultmann were all deficient in their views. What we do know is that Israel saw the events of their history as having the form of promise and covenant. The exodus from Egypt was the work of Yahweh who was not only the Lord of history but the Creator who transcended and entered creatively into it. This God 'went before' them, 'luring' them on into the future. This eschatologial element in the faith of Israel was spoken of by prophet, seer and people, as they looked forward to an end-time, a new age when God would be all in all.

There is a radical novelty in such a future yet it is related to past and present. The promises made to the patriarchs become surprisingly new in the fulfilment of the exodus so the history of Israel revealed an openness to the future. The idea of a local God involved in their history became the Yahweh who controls the destiny of the nations. The experience which they had of the world empires from the period of the exile onwards made the sovereignty of God explicit and was expressed strongly in their apocalyptic writing: the meaning of history will only be fully known at its end. W. Pannenberg holds that we cannot present Jesus without this context otherwise we go back to the nineteenth-century lives of Jesus, which endeavoured to see him through the eyes of modern Europeans.

The same element was present in the teaching of Jesus and though there is much debate about the sayings of Jesus and the titles that may have been ascribed to him during his earthly life there is a body of opinion which holds that Jesus did make a claim to authority and a demand for decision in relation to his person and message. He claimed to usher in 'the end-time'.[22] Was Jesus mistaken? Pannenberg believes he was not and argues that with the resurrection the 'end-time' was anticipated in his person. It was a foretaste of the end of history which corresponds to the Hebrew view of truth found in history and revealing itself in the future. The resurrection confirms that Jesus is the beginning of the final eschatological event.

Pannenberg's perspective, then, rests on the apocalyptic view of history and the hope that death will not end everything. If death is final it makes nonsense of all promise and fulfilment. The apocalyptic, though clothed in symbols, expresses the universal context of hope and faith. For the individual, no matter how long life is, death comes too soon for there is so much to do. If the religious framework is not acceptable for hope to be maintained we may opt for a belief in progress or socialism or some form of communism or whatever. Man's hope accepts either a secular or a religious structure for without it he cannot live. Biblical eschatology is not so far away from modern thought as some secularists think, for science sees the end of the world in a far more bizarre and dramatic way than anything that the Hebrew imagination could conceive. The book of Revelation is 'a pale document compared with these modern scientific apocalypses'.[23] The difference is the gloom of the scientific picture as compared with the hope of the biblical one.

It is generally accepted that Jesus preached the kingdom of God, which was present but yet future. The Gospels were written after the resurrection, that is the end of the event, and it is from this perspective that the writers understood who and what he was. For them the resurrection confirmed not only his eschatological message but himself as the *eschaton*: the beginning of the end of history. Pannenberg draws out implications for Christology which I discuss elsewhere[24] but with regard to the apocalyptic he is making the point that the full meaning of history will be known only at its end. Just as we only know the real significance of an event or person at the end the same applies both to Jesus (revealed as Son of God by the resurrection) and the end of the world. Only then will the full justice and glory of God be disclosed. We recall that the same applied to suffering.

But can we believe in the apocalyptic vision of the end of the world, final judgement, general resurrection of the dead in the body, and the establishment of a new heaven and a new earth? Bultmann could not, and de-mythologised the scheme and produced his existentialist interpretation. Pannenberg is fully aware of his work and the later historical criticism of the Gospels which we have referred to in a previous chapter, so he minimises the amount of information required for his position. All he needs is that Jesus did make a claim to authority in relation to his person and message which did not become clear until after his resurrection. There is no attempt to establish the divinity of Christ on the basis of his pre-Easter history, which literary criticism has been unable to settle.[25] With regard to such a claim even Bultmann concedes that Jesus did demand a decision for his person as the bearer of the Word and post-Bultmann writers confirm this. Pannenberg recognises that Schweitzer was right in maintaining the apocalyptic message of Jesus but incorrect as seeing it as something foreign to our way of thinking. Only its symbolic form is strange to us, not its universal validity based on our openness to that which is beyond this world.

John Hick for his part holds that there will be a life after death and an embodied one and cites the parable of Dives and Lazarus and the sheep and the goats. God is faithful and will not allow those who have sought to realise the highest potentialities of their nature to perish. Death will not cancel God's love for us. He questions a merging of our consciousness in the divine since it means the loss of individual personality, a drop returning to the ocean. We would agree that there is a need for individual personal existence or identity and have argued that the laboratory of the world is for the making of souls. If this is so then we require a life beyond this one, where injustices will be dealt with and the good rewarded. We do not realise our potentialities before we die; life after death would provide an opportunity for further spiritual growth. But the view leads Hick into speculation about possible reincarnation or rebirth perhaps not only here but in other worlds.[26]

As we have said, body and soul are together in Christian thought but the human has been animated by the breath of

God (Gn. 2:7) and so shares in the image of God. The soul is the true essence of a person and is the centre of the inner life, the place of feelings and emotions (1 Thess. 2:8). It is an important possession (Mk 8:36–7 and overlaps with *pneuma* or spirit. The flesh is the location of sin but the soul and spirit are for salvation and participation in the divine.[27] The spiritual quality that we possess yearns for a life beyond our present existence, one with a new body not subject to the temptations of this one.

As we mentioned, Immanuel Kant put forward the moral argument for life after death. If God exists, the highest good will finally be realised, not here but hereafter. Kant believed in an objective moral law but today so much is relative and such a law has been questioned. Morality, it is said, is a matter of feeling and personal opinion and emerged to hold societies together. While this is true many would contend for some objective quality about morality which makes us question any ethics based on the survival of the fittest. Philosophers also ask whether life after death is desirable.[28] But the believer, and it might be thought most people, considering the injustices of this life, would hold that such a life free from suffering is very desirable.

In conclusion, we can emphasise that hope of immortality helps us to be patient in the test of suffering, is somehow ingrained in the human race and proclaimed by most religious traditions from the beginning.

9
Objections to the Model of the Divine Scientist

In this final chapter I would like to try and answer some possible objections to the model of the divine scientist which I think might arise. Readers of the book hopefully will let me have others that occur to them. It has always seemed to me that in such dialogue there is the possibility of making progress.

Does the model not depart too much from traditional ideas?

We have noted the usual models for God, that is, father, mother, husband, shepherd, friend, and so on. They remain popular not only for the sense of intimacy which they impart but because they are scriptural. Others have arisen with the advance of science: divine clockmaker, workman-tool (primary cause working through secondary causes), determiner of indeterminacies, communicator of information, agent–action, and leader–community.[1] Looking at these, the model of divine scientist does not seem to be out of place but a continuation of the effect of a scientific age.

But we have also argued that it can arise from looking at the design and purpose in the world. David Hume was critical of the argument from design but did admit that an organising principle responsible for patterns in nature might be within organisms not external to them. Today we see God allowing nature to make itself through a long history of evolution which means that design is built in. If there is design then there is purpose, admitted by Dawkins, even if he restricts it to man, but Darwin said that he could not think of the totality of the world as the result of chance. His problem, as we said, was seeing each separate thing as the result of design. Some physicists see the universe coming into existence because of the quantum laws: it 'fluctuates' into existence but such laws do not allow us to call it chance. The odds against it are immense.[2] Where the laws came from is still debatable, with some thinking of them as eternal and others as invented by us or whatever. But it is the rationality of such laws which can be expressed in mathematics that point to a designer.

Perhaps we are right in thinking not only from design but to it. Evolution places the stress on development and the theist sees things proceeding to a goal. The most significant stage in the evolutionary process has been the change from instinctive behaviour to that which was a conscious response to value. God is influencing us towards the values of goodness and love.

But a mystery surrounds the beginning of life. Paul Davies says that the creating of life in a test tube remains a distant dream and there is little support in the belief that life should arise inevitably, given earth-like conditions. Making the building blocks of life in a test tube is theoretically possible and amino acids have been found in meteorites, but just as bricks alone don't make a house it takes more than a random collection of amino acids to make life. They must be linked together in long chains in the right order. In a living organism energy is fed into the system by the cell's molecular machinery with its intricate specifications. There is a logical structure and organisation of the molecules. The DNA which we discussed is the genetic databank and the genes are instructions for making proteins and other molecules. Hence we have an information-processing system which organises complexity and Davies goes on to point out that it is the information content or software of the cell that is the mystery.

How did atoms write their own software? Where did the genetic information required come from? Biological information is not encoded in the laws of chemistry or physics; Darwinism shows how organisms acquire information, but this is when life is already under way. In reply molecular Darwinism points to replicating molecules in some sort of chemical soup and that they were formed by chance. But the DNA would hardly form by chance and this is true also of RNA. Altogether there is too much required of chance, for nature must provide replicators simple enough to form by it, able to replicate accurately and with a huge range of variants for selection to act upon.[3]

We also discussed body and mind in Chapter 7 and if the human mind transcends the brain then analogously we can think of a cosmic mind or self which transcends the universe. It is the rational principle or *logos* (Stoics) or sum of the divine ideas (Philo) or Christ as the eternal Word (John 1:1). Some see the forms of Plato in the structure of the physical world. Thus the continuing conformity of physical particles to precise mathematical relationships is something that is much more likely to exist if there is an ordering cosmic mathematician who set up the correlation in the requisite way.[4] Augustine thought that the creator was the divine geometer and the perfection of the divine order was seen in the patterns that permeate the universe. When we consider the vast cosmos, with its 2,000,000 stars and about 100,000 million galaxies operating with such precision it reflects the planning of a scientific cosmic mind. It would seem that this points to a divine experiment taking place in the laboratory of the world. God controls the experiment, using 'chance' to explore its possibilities and 'necessity', expressed in the natural order of the world, setting limits to it. But it involved risks as shown in the fall of humanity. Stephen Hawking says God not only plays dice but throws them!

Does the model fail to stress the personal relationship which exists between us and God?

Only if it is seen as excluding other models, such as we mention above. These, particularly father and mother, are more useful in the context of prayer and worship. Only if the model of the scientist is not held in tandem with other images such as these would it have difficulty. They are very intimate but the Psalmist holds the model of father together with the model of the cosmic God as he surveys the heavens. The idea of the cosmic scientist fits in with his transcendence whereas the more familiar and intimate ones agree more with his immanence. In Genesis 1 we have the more abstract and impersonal picture of the creator who just speaks the Word of creation and it is done but in Genesis 2 we have the anthropomorphic and personal. In the New Testament we have the *logos*, the rational principle behind all things and the personal incarnation in Christ.

God can be compared to a scientist but only if the caricature of the human scientist depicted in fiction and films as clinical, rational, impersonal, objective and devoid of personal feelings, is set aside. The rationality is there but also the personal elements of compassion, judgement and creative imagination. The scientist cannot escape the values embraced by the religions and our belief in morality. Knowledge as shown in the social sciences is often gained by participation and in the Christian scheme it is evident in the ministry of Christ. As we said, the social scientist is a more appropriate analogy than the physicist, though the observer plays a crucial role in the understanding of quantum mechanics. The biologist too can establish an intimate relation with what she studies, as is shown by work done with chimpanzees which respects their human qualities. It is through interaction that knowledge is obtained and the same applies in connection with God. He is known by what he does.

Does the model not depart from the traditional view of the power of God?

It does, but only because we have argued that some form of *kenosis* is necessary due to creation. Because of this it is argued that we can apply limitation to God but it is a self-limitation and God retains his power to be used when necessary, as in the resurrection of Christ and the final consummation of all things.

Does it help in understanding the problem of evil?

The basic answer is that if God is a scientist then an experiment is being carried out and it will involve testing. We noted the order and disorder in the world with the former reflected in the laws and the latter arising because of evil. With regard to it, in Chapter 6, we referred to life as a test. From childhood we experience it in relation to home, education, employment, and the various roles we play. Often the Darwinian principle of the survival of the fittest operates. But we react against the evolutionary process and seek values which we feel ought to govern our lives, hence we try to achieve justice, mercy and goodness. At the higher level, history is dotted with saints and martyrs who were willing to die for their beliefs. In other words, while recognising the need for survival and reproduction we commit ourselves to higher goals because we think these 'ought' to be. Our characters are tested throughout life and are refined by the process. Evolution itself looks like a vast experiment with continuous and flexible creativity and it reminds us that pain is necessary, for new forms arise only when the old die.

But it was acknowledged that testing is only one answer to the problem of evil for while many say God was with them others complain of his absence. For many the price of suffering is too high and they complain of the absence of God. Others, however, like the apostle Paul, say that they received grace to bear the absence and could rejoice even in suffering. Attitudes to suffering extend from resignation to rebellion and to the joy of participating in the suffering of Christ. But it is natural to rebel under the stinging lash of affliction. Where is the justice involved when the suffering is undeserved? We have noted in particular the case of Job who spoke of the bitterness of his soul but soon realised that this negative attitude would not answer his problems. To go on nursing a grudge against how life has treated us is to prevent the perceiving of a reason for it. Or we can resign ourselves to it as Eli, the High Priest of Israel, did when hearing of the tragic death of his sons, 'It is the Lord: let him do what seems good to him' (1 Sam. 3:18) or Job who, advised by his wife to curse God and die, countered, 'What! shall we receive good at the hand of God, and shall we not receive evil?' (2:10) Or the Psalmist, 'I was dumb, I opened not my mouth; because you did it' (38:13). In many ways it is a good attitude so that even in some awful bereavement we say, 'your will be done'. It was the cry that Jesus uttered in Gethsemane while he sweated drops of blood (Lk. 22:44).

But it is very difficult to reach the position of rejoicing as with Paul. It is showing sheer grit, a stiff upper lip, and a stoical smile but more than that, for it looks like a counsel of perfection, rejoicing in suffering and over it. Yet we see the practical nature of it when the record speaks of Paul and Silas, beaten, bruised and imprisoned, singing songs in the night (Acts 16:25). Paul, later in life, was imprisoned in a dungeon, aged, lonely, infirm and almost blind, and knew that at any moment he could face execution and his missionary dream would be over. But he rejoiced because he had hope of life after death, 'for our light affliction which is but for a moment, works for us a far more exceeding and eternal weight of glory; while we look not at the things which are seen but at the things which are not seen; for the things which are seen are temporal but the things which are not seen are eternal' (2 Cor. 4:17, 18).

Suffering too, as we have seen, provides moral insight in a way that pleasure cannot; it teaches the value of discipline, faith and hope. Christian biography is full of examples of how suffering can be transformed into sympathy and service. It provides the opportunity of doing good as we saw in the herculean efforts made by the medical profession to save life. As they battle with disease we are reminded of Jesus' ministry of healing, and the Christian recognises that he must share in the suffering of Christ. With him was the initial defeat of evil. In connection with moral evil God does not intend it but opens up the possibility of our rejecting it. Hence the test.

It would seem that testing requires the presence of both good and evil and we tried to say something about the origin of evil. Evolution does help in this respect, for defective design can be explained as due to natural causes. Those who argue that it should not be, need to show what kind of world they would create and how it would produce the values of love and obedience. But the greatest test is the suffering of the innocent and the idea of testing has difficulty here with regard to many cases. Yet Christ was innocent and he was able to turn his suffering into benefit for humanity.

Is too much freedom given to mankind?

We noted that it is the freewill defence which is often used when the problem of moral evil is discussed. Would it have been possible to limit such freedom so that evil would not arise? The difficulty is that love requires a free response and God will not coerce us to obey him. The alternative was to make us with robot-like qualities, but what would be the point of a creature predetermined and programmed to obey? We love those who freely appreciate what we do and say and enter into a loving relation with them. Free will is a dangerous quality to place in our hands but God apparently thought the risk was worthwhile.

Does it help in understanding the action of God?

With regard to the world, the Newtonian machine model has been replaced by a web of religionships stretching down to the sub-atomic level so that one thing cannot be understood without reference to another. Religion and influence replace rigid cause and effect. It was once thought that anything which influences the position of an object would have to have some sort of physical force, but in quantum theory an electron in a probability distribution does not even have a definite position.

What has been called 'quantum entanglement', the link between two or more photons, electrons or atoms, even if they inhabit distant parts of the universe, has puzzled physicists for many years. When one is measured it affects the other and it happens instantaneously even if they are widely separated. The link of entanglement knows no boundaries, as Mark Buchanan points out; it isn't a cord running through space but lives somehow outside of space. It goes through walls and pays no attention to distance. But how can this happen since nothing travels faster than light? One answer is that both relativity and quantum theory must be combined. The first protects the chain of cause and effect and ensures that effects never happen before their causes. Quantum theory stops just short of upsetting this. Relativity ensures a degree of separateness and individuality for distant pieces of the universe while quantum entanglement keeps the whole universe coherently connected. Other scientists try to advance beyond quantum theory, trying to find some other principle where non-locality and causality can exist together.[5] But there does seem to be a plan in all of this, pointing again to the Mind behind the universe.

What is of interest to us is that such influence between sub-atomic particles exists. We know on the higher levels of persons how one will affect another in the choice of a career, morality, marriage, and so on. We cannot measure influence as we do physical things. How can ideas, feeling and personal relationships be quantified? God can be seen as the experimenter responding and initiating new courses of action and attracting the development of positive potentialities. He does not force but through parents, scripture, church, friends, acts to bring us into right relationship with him.

In concluding this book, we recall that religions need not only to conserve but to accommodate to society if they are to communicate their message. Otherwise they tend to lapse into a ghetto mentality. We noted that the writer of the Fourth Gospel, with his *logos* model, tried to communicate with the Greeks of the day and the same was true of the Apostle Paul in his speech to the intelligentsia at Athens. His approach was courteous and started with their ideas about the gods or 'the unknown God' (Acts 17:23) and then went on to try and enlighten them regarding the true God. He proclaimed the resurrection and called on them to repent of their ignorance. As usual when the resurrection was mentioned it was treated as ludicrous or with sneers or the promise to consider such matters later (vs. 32), but some were convinced. While the speech has been debated, some saying it was a failure while others think it was a success, it is significant for Paul's attempt at natural theology shown by quotations from two pagan authors. It would appear that for him glimmerings of truth and insight from such general revelation could be found in non-Christian authors. He was attempting to find a point of sympathy and contact with them in order to build a bridgehead of persuasion.[6] Today around the Parthenon in twenty-first century Athens there is a street called 'Apostle Paul Street' and on Mars Hill is a bronze tablet recording the sermon as summarised by Luke. A testimony to its lasting effect.

We also have tried to make contact with our age with the model of God as the scientist and we hope it will persuade some at least of its relevance to our age. Of course, as we have mentioned, all our models of God are inadequate and we recognise the limitation of this one. Hence it seems appropriate to end with what Hilary of Poitiers said:

> There can be no comparison between God and earthly things but the weakness of our understanding forces us to seek certain images from a lower level. Hence every comparison is to be regarded as helpful to men rather than suited to God since it suggests rather than exhausts the meaning we seek.[7]

Notes

Introduction

1. A. Einstein, *Ideas and Opinions*, Souvenir Press, London, 1973, p. 46.

1 The New Method of Science

1. I. Barbour, *Religion and Science*, HarperCollins, New York, 1997, p. 106.
2. J. Polkinghorne, *Searching for Truth*, SPCK, London, 1996, p. 15.
3. M. Fuller, *Atoms and Icons*, Mowbray, London, 1995, p. 18.
4. A.F. Chalmers, *What is This Thing called Science?* Oxford University Press, Milton Keynes, 1978, pp. 38f, 54. See Karl Popper, *Logic of Scientific Discovery*, Hutchinson, London, 1968; *Conjectures and Refutations*, Routledge & Kegan Paul, London, 1969.
5. T.S. Kuhn, *Structure of Scientific Revolutions*, University of Chicago Press, Chicago, 1970, pp. 147f.
6. I. Barbour, op. cit., p. 127, M. Fuller, op. cit., p. 21
7. P. Luscombe, *Groundwork of Science and Religion*, Epworth Press, London, 2000, pp. 75f.
8. A.F. Chalmers, op. cit., p. 104.
9. A.F. Chalmers, op. cit., pp. 124, 139, 141.
10. A.F. Chalmers, op. cit., pp. 74ff.
11. P. Luscombe, op. cit., p. 84, M. Fuller, op. cit., p. 23.
12. J. Hawking, *Music to Move the Stars: A Life with Stephen Hawking*, Pan, London, 2000, p. 465.
13. A.F. Chalmers, op. cit., p. 31.
14. A.F. Chalmers, op. cit., p. 46.
15. I. Barbour, op. cit., p. 110.
16. D. Brian, *Einstein: A life*, John Wiley and Sons, New York, 1996, p. 200.
17. J.D. Barrow, 'Frontiers and Limits of Science' in *How Large is God*, ed. by John Marks Templeton, Templeton Foundation Press, Philadelphia and London, 1997, p. 211.
18. P. Luscombe, op. cit., 102ff, 118ff.
19. R.T. Wright, *Biology through the Eyes of Faith*, Apollos, 1991, pp. 41f.

2 Religion and Scientific Method

1. M. Poole, *A Guide to Science and Belief*, Lion, Oxford, 1994 edn, p. 24.
2. J.W. van Huyssteen, *Duet or Duel? Theology and Science in a Postmodern World*, SCM, London, 1998, pp. 15f.

3. I. Barbour, op. cit., pp. 134, 266.
4. J. Gribbin and M. Gribbin, *Almost Every Mans's Guide to Science*, Phoenix, London, 1999, p. 94. M. Fuller, op. cit., p. 25.
5. W.J. Abraham, 'Revelation Reaffirmed' in Paul Avis, ed., *Divine Revelation*, Darton, Longman and Todd, London, 1997, pp. 206f.
6. J. Polkinghorne, *Quarks, Chaos and Christianity*, Triangle, SPCK, London, 1994, p. 9.
7. M. Polanyi, *Personal Knowledge: Towards a Post Critical Philosophy*, Harper and Row, New York, 1964, passim.
8. T. Torrance, *Preaching Christ Today*, W.B. Erdmans Publishing Co., Michigan, 1994, pp. 2ff.
9. S.S. Smalley, 'The Gospel according to John', *The Oxford Companion to the Bible*, ed. by B.M. Metzer and M.D. Coogan, Oxford University Press, Oxford, 1993, p. 374.
10. T. Torrance, op. cit., p. 6.
11. T. Torrance, op. cit., p. 24.
12. J. Polkinghorne, *Scientists as Theologians*, SPCK, London, 1996, p. 64.
13. D. Migliore, *Faith Seeking Understanding: An Introduction to Christian Theology*, W.B. Eerdmans, Michigan, 1991, pp. 19ff.
14. M. Poole, op. cit, p. 59.

3 God Talk

1. M. Poole, op. cit., p. 77.
2. G.B. Caird, *Language and Imagery of the Bible*, Duckworth, London, 1980, pp. 47f.
3. M. Poole, op. cit., p. 14.
4. G.B. Caird, op. cit., p. 174.
5. G.B. Caird, op. cit., pp. 32, 145.
6. G.B. Caird, op. cit., pp. 193, 197.
7. M. Poole, op. cit., p. 75. John Houghton, *The Search for God: Can Science Help?* Lion, Oxford, 1995, p. 122.
8. I. Barbour, op. cit., p. 118.
9. A McGrath, *Glimpsing the Face of God: The Search for Meaning in the Universe*, Lion, Oxford, 2002, p. 79.
10. R. Penrose, *The Emperor's New Mind* Oxford University Press, Oxford, 1989, p. 146. P. Davies, *The Mind of God*, Simon and Schuster, London, 1992, p. 84, quoted by J. Houghton, op. cit., p. 203.
11. K. Ward, *God, Faith and the New Millennium*, pp. 16f. One World, Oxford, 1998.
12. A. Beardslee, 'Logos', *Oxford Companion to the Bible* as cited above, p. 463.
13. H.W. Bartsch, ed., *Kerygma and Myth*, vol. 1, SPCK, London, 1953, p. 10, note 2.
14. J. Hick, *God and the Universe of Faiths*, p. 167, Macmillan, London, 1973.
15. Ibid., p. 230.

16. J. Hick, *The Fifth Dimension*, pp. 235ff, One World, Oxford 1999.
17. R.H. Stein, 'Parables', *Oxford Companion to the Bible*, as cited above, p. 568.
18. Ibid., J. Johnsson, 'Irony and humour', pp. 324f.
19. A. Richardson, *Genesis 1–11*, SCM, London, 1963, pp. 27ff.
20. R. Crawford, *The God/Man/World Triangle*, Macmillan, London, 2000, p. 189.
21. R. Crawford, ibid., p. 190. See D.Z. Phillips, *The Concept of Prayer*, London, 1965, Ludwig Wittengstein, *Philosophical Investigations*, Basil Blackwell, Oxford, 1953; G.E.M. Anscombe, Oxford University Press, Oxford, 1968.

4 The Experiment Begins

1. M. White and J. Gribbin, *Stephen Hawking: A Life in Science*, Penguin, Harmondsworth, 1998.
2. R. Stannard, *100 Questions Part 2*, pp. 140, 197, Faber and Faber, London, 1998. M. Fuller, op. cit., p. 81.
3. J.D. Barrow, *The Origin of the Universe*, Orion, London, 1995, p. 27.
4. I. Stewart, *Does God Play Dice?* Blackwell, Oxford, 1989, pp. 2f.
5. P. Davies, ed., *Superstrings: A Theory of Everything?* Cambridge University Press, Cambridge, 1988, p. 5.
6. Thanu Padmanabhan, *After the First Three Minutes: The Story of Our Universe*, Cambridge University Press, Cambridge, 1998, passim. See also 'Quantum Cosmology and the Creation' in *Frontiers of Science*, ed. A Scott, Blackwell, Oxford, 1990, pp. 153ff.
7. Quoted by L. Weatherhead, *The Christian Agnostic*, Hodder & Stoughton, London, 1965.
8. B. Davies, *Introduction to the Philosophy of Religion*, Oxford University Press, Oxford, 1993, pp. 101ff.
9. P. Davies, *The Edge of Infinity*, p. 171, Penguin, London, 1994. Van Huysteen, op. cit., pp. 67ff.
10. M. Poole, op. cit., p. 116.
11. M. Poole, op. cit., p. 119.
12. J.W. Van Huyssteen, 'Theology and Science, the Quest of a New Apologetics', *Princeton Seminary Bulletin*, No. 2, 1993, pp. 113f.
13. M.D. Coogan, 'Chaos' *Oxford Companion*, op. cit., p. 105. J. R. Porter, 'Creation', op. cit., p. 140.
14. G.B. Caird, op. cit., footnote p. 237.
15. J. Polkinghorne, ed., *The Work of Love: Creation as Kenosis*, SPCK, London, p. 2001.
16. J. Polkinghorne, ibid., pp. 194–9.
17. I. Barbour, 'God's Power: A Process View', ibid., pp. 7ff.
18. I. Barbour, ibid., p. 12.
19. A.R. Peacocke, ibid., pp. 37, p. xi.
20. A.R. Peacocke, ibid., pp. 224–41.

21. J. Moltmann, ibid., pp. 200–44, H. Rolston, 'Kenosis and Nature', ibid., p. xi.
22. J. Polkinghorne, ibid., p. 105.
23. P. Vardy, *The Puzzle of Evil,* Fount, HarperCollins, London, 1992, pp. 147, 159–65.

5 The Subjects of the Experiment

1. M. White and J. Gribbin, *Darwin: a Life in Science,* Simon and Schuster, London, 1995, p. 115.
2. R. Crawford, op. cit., p. 36.
3. L. Gamlin, *Evolution,* Dorling Kindersley, London, 1996, pp. 58, 143.
4. R. Snedden, *Genetics,* Wayland, Hove, 1995, p. 35.
5. J. Bowker, *Is God a Virus*? SPCK, London, 1995 pp. 22f, 47.
6. J. Bowker, ibid., pp. 56, 72.
7. K. Ward, op. cit., p. 112.
8. K. Ward, op. cit., pp. 111, 116.
9. J.W. van Huyssteen, *Duet or Duel? Theology and Science in a Postmodern World,* SCM, London, 1998, p. 127.
10. I. Barbour, op. cit., p. 223.
11. P. Vardy, op. cit., pp. 103–6.
12. G. Dover, *Dear Mr Darwin,* Weidenfeld & Nicolson, London, 2000, pp. 61ff.
13. G. Dover, ibid., p. 202.
14. G. Dover, ibid. Steve Rose calls it 'genetic imperialism'. See his *Lifelines: Biology, Freedom, Determinism,* Penguin Press, London, 1997.
15. P. Davies, *The Cosmic Blueprint,* Heinemann, London, 1987, pp. 102ff.
16. L. Spinney, 'The unselfish gene', *New Scientist,* London, 25 Oct. 1997, p. 32.
17. G. Dover, op. cit., pp. 65, 76.
18. A. R. Peacocke, *Creation and the World of Science,* Clarendon Press, Oxford 1979, ch. 3, and *Theology for a Scientific Age,* Fortress Press, Minneapolis, 1993, ch. 9.
19. G.B. Caird, op. cit., pp. 135–6.
20. A.R. Peacocke, op. cit., p. 193.
21. R. Stannard, op. cit., pp. 58ff.
22. J. Hick, *God and the Universe of Faiths,* pp. 45, 97f. Macmillan, London, 1973.
23. A.R. Peacocke, op. cit., p. 196.
24. K. Ward, op. cit., pp. 109ff., 123–7. See also *In Defence of the Soul,* One World, Oxford 1998.

6 The Test

1. P. Vardy, *The Puzzle of Evil,* pp. 45, 111f., HarperCollins, London, 1992.
2. G.B. Caird, op. cit. p. 225.

3. A. Plantinga, *God, Evil and the Metaphysics of Freedom*, Oxford University Press, Oxford, 1990, p. 108.
4. D.A.S. Fergusson, *The Cosmos and the Creator*, SPCK, London, 1998, p. 23.
5. W. Barclay, *Gospel of Matthew, Vol 2*, chs 11–28, St Andrew Press, Edinburgh, rev. edn, 1975, p. 368.
6. J. Hick, *Evil and the God of Love*, Harper & Row, New York, 1966, p. 285.
7. J. Hick, ibid., p. 176.
8. J. Hick, ibid., p. 237.
9. J. Hick, *God and the Universe of Faiths*, p. 74.
10. P. Vardy, op. cit, p. 22.
11. F. J. Ayala, ' "Intelligent design": the original version', pp. 9ff., in *Theology and Science*, vol. 1, no. 1, 2003, Routledge, London.
12. R. Stanard, ed., *God for the 21st Century*, SPCK, London, 2000, p. 31.
13. K. Kohler 'A priest people' (p. 64), Louis Jacobs, 'The chosen people' (pp. 65f), in *The World Religions Reader*, ed. by G. Beckerlegge, Routledge, London, 1998.
14. S. Kolitz, 'Y. Rakover's Appeal to God', Beckerlegge, op. cit., pp. 48f.
15. J. Hawking, op. cit., pp. 461, 572, 596.
16. I. F. Zygmuntowicz with Sara Horowitz, 'Survival and Memory', Beckerlegge, op. cit., pp. 51f.
17. J. Hick, *Evil and the God of Love*, Harper & Row, New York, 1966, p. 257.
18. R. Crawford, op. cit., p. 208.
19. A. Walker, *The Many-Sided Cross of Jesus*, Epworth Press, London, 1962, p. 59.
20. R. Stanard, *The God Experiment*, pp. 92f., Faber and Faber, London, 1999.

7 The Action of the Divine Scientist

1. J. Polkinghorne, *Scientists as Theologians*, SPCK, London, 1996, pp. 9, 34.
2. J. Polkinghorne, ibid., pp. 36f.
3. J. Polkinghorne, ibid., pp. 71.
4. P. Davies, 'Life force', *New Scientist*, 18 Sept. 1999, pp. 27ff.
5. R. Crawford, op. cit., pp. 95ff.
6. A. Gellatly and O. Zarate, *Mind and Brain*, Icon, Cambridge, 1998, p. 11.
7. E.O. Wilson, *On Human Nature*, Harvard University Press, Cambridge MA, 1978, passim.
8. R. Penrose, *Shadows of the Mind*, Vintage, London, 1994, pp. 19, 45, 52f.
9. J. Houghton, op. cit., pp. 97ff.
10. S. Rose, 'Minds, Brains and Rosetta', pp. 201ff., *New Scientist*, 25 Oct. 1997.
11. I. Barbour, *Issues in Science and Religion*, SCM, London, pp. 443f.
12. G.B. Caird, op. cit., p. 79.
13. K. Ward, op. cit., pp. 175ff.
14. Leslie Weatherhead, op. cit., p. 26.
15. R.H. Fuller, 'Jesus Christ, Life and Teaching': critical method, *Oxford Companion*, op. cit., pp. 357ff.

16. R.H. Fuller, ibid., p. 363.
17. W. Beardslee, 'Incarnation', *Oxford Companion*, op. cit, p. 301.
18. W. Barclay, op. cit., p. 83.

8 The End of the Experiment

1. M. Hanlon, *Daily Mail*, 22 Aug. 2003, p. 13.
2. R. Matthews, 'To infinity and beyond', *New Scientist*, 11 April 1998, pp. 27ff.
3. F. Tipler, *The Physics of Immortality*, Macmillan, London, 1995, pp. 124f.
4. A.J. Klotzko, 'There'll never be another you', *Guardian*, 22 Jan. 2004. See also *A Clone of Your Own*, Oxford University Press, Oxford, 2004.
5. M. Buchanan, 'Beyond reality', *New Scientist*, 14 March 1998, pp. 27ff.
6. B. Davies, op. cit., p. 231.
7. S.A. White, 'Afterlife and immortality', *Oxford Companion*, op. cit, 1993, p. 17.
8. D. Cohn-Sherbok, 'The Philadelphia Platform of Reform Judaism', *Holocaust Theology: A Reader*, New York University Press, 2002, p. 136.
9. See R. Crawford, *What is Religion?*, Routledge, London, 2002, for more on Hinduism.
10. R.C. Zaehner, *The Bhagavadgita*, Oxford University Press, Oxford, 1969, pp. 10f.
11. See R. Crawford, op. cit., for more on Buddhism.
12. S.H. Travis, 'Second Coming of Christ', *Oxford Companion*, op. cit., p. 685.
13. J. Polkinghorne, *The God of Hope and the End of World*, SPCK, London, 2002, pp. 95, 233.
14. J. Hick, *Evil and the God of Love*, pp. 344f.
15. J. Baker, *The Foolishness of God*, Darton, Longman and Todd, London, pp. 252f.
16. S.T. Davis, D Kendall, G. O'Collins, eds, *The Resurrection*, Oxford University Press, Oxford, 1997, p. vii.
17. Ibid., pp. 6ff, 33.
18. Ibid., pp. 142, 156, 176, 182. R. H. Fuller, 'Resurrection of Christ', *Oxford Companion*, op. cit., pp. 647f.
19. M. Fuller, op. cit., p. 93.
20. J. Polkinghorne, op. cit., pp. 56, 76, 107, 117, 243.
21. K. Ward, op. cit., pp. 190ff.
22. W. Pannenberg, *Jesus, God and Man*, SCM, London, 1980, p. 100.
23. A.D. Galloway, *Wolfhart Pannenberg*, George Allen & Unwin, London, 1973, p. 51.
24. R. Crawford, *Saga of God Incarnate*, University of South Africa, Pretoria, second edition, 1988, p. 49.
25. A.D. Galloway, op. cit., p. 65, R. Bultmann, *Theology of the New Testament*, SCM, London, 1971, Vol, i, p. 43.

26. J. Hick, *God and the Universe of Faiths*, pp. 185f. See also *The Fifth Dimension: An Explanation of the Spiritual Realm*, One World, Oxford, 1999, p. 245.
27. S. White, 'Human Person', *Oxford Companion*, op. cit., p. 296.
28. B. Davies, op. cit., p. 231.

9 Objections to the Model of the Divine Scientist

1. I. Barbour, *Religion and Science*, Harper, San Francisco 1997, p. 305.
2. R. Crawford, *God, Man, World Triangle*, for a more detailed discussion, pp. 201ff.
3. P. Davies, 'Life Force', *New Scientist*, 18 Sept 1999, pp. 27ff.
4. R. Crawford, op. cit., p. 203.
5. M. Buchanan, 'Why God plays dice', *New Scientist*, 22 Aug. 1998, pp. 27ff.
6. D. Clarke, 'A Sense of Place', *Herald*, Presbyterian Church, Belfast, March 2004.
7. Quoted in R. Crawford, op. cit., p. 154. See the *New Catholic Encyclopedia*, vol. 6, 1967, McGraw-Hill, New York, p. 1114. For Hilary (AD 315–75) (Bishop of Poitiers), the nature of God is incomprehensible. Quoted also by John Houghton, *The Search for God*, Lion, Oxford, 1995, p. 128.

Index

Abraham, 69, 95
Abraham, W.J., 165
Absence of God, 97–8
Act of God, 5, ch. 7
Adam, 87–8
Adler, 10
Aggregates, 139
Alexandrian school, 51
Allah, 38
Allegory, 51
Alston, William, 145–6
Altruism, 86
Amos, 38
Analogy, 41–2, 145
Anselm, 1, 35
Anthropic principle, 58–60
Apocalyptic, 4, Ch. 8
Aquinas, Thomas, 1, 35, 99
Archetype, 88
Argument from design, 63
Arhats, 140
Aristotle, 76
Arjuna, 28, 96
Athens, 162
Atkins, Peter, 64
Atonement of Christ, 88
Augustine, 62, 98, 157
Augustinian, 99
Auxiliary hypothesis, 25, 35
Ayala, F.J., 169

Barbour, Ian, 11, 15, 25, 42, 60–70, 82, 111, 165–71
Barrow, John, 60, 167
Barth, Karl, 26, 29, 31, 88, 144
Bartsch, H.W., 166
Beardslee, A., 166, 169
Behaviourism, 114
Benzine ring, 25
Bhagavadgita, 28, 96, 139
Bible, 37
Biblical eschatology, 141–3, 151
Big Bang, 58–61, 65, 133
Big Crunch, 134
Black holes, 16, 31, 58
Blackmore, Susan, 135
Blasphemy, 50, 68, 130
Bodhisattva, 47
Bohm, David, 17–18
Bohr, Nils, 16–18
Book of Revelation, 141, 151
Bottom-up, 113
Bowker, John, 79, 168
Brain, 115–18
Brahman, 27, 47, 105
Brian, D., 165
Brunner, Emil, 28
Buchanan, Mark, 161
Buddha, 34, 47–8, 96, 140
Buddhism, 24, 34, 139
Bultmann, R., 40, 49, 125, 147, 150, 152

Caesar, 144
Caird, G.B., 40, 166–9
Calvin, John, 98
Catholicism, 24
Cavendish Laboratory, 23
Chalmers, A.F., 13–14, 165
Chambers, Robert, 76
Chance, 63–4, 85
Chaos, 66
Chaos theory, 60–1, 124
Christ, 27, 39–40, 46–7, 68, 97, 152
Chromosomes, 78–80
Church, 40
Cloning, 135
Cobb, John, 121
Cohn-Sherbok, D., 138, 170
Community, 3, 11, 19, 25, 32

Complement, 4
Computers, 116, 118
Confessions of faith, 35
Consciousness, 115–18
Coogan, M.D., 167
Copernicus, 1, 12, 63
Cosmic mind, 64, 157
Cosmic scientist, 45, 93
Covenant, 69
Crawford, Robert, 167–71
Creation, 65–7, 95
Creative principle, 43
Crick, Francis, 77–8
Culture, 81
Cupitt, Don, 53–5
Curie, Marie and Pierre, 107
Cyclic time, 137

Daniel, 103
Dark matter, 24, 60
Darwin, Charles, 7, 9, 14, 20, 25, 42, 64, 75–84, 156
Darwinism, 157
Data of religion, 3, 24
David, 51
Davies, B., 170
Davies, Paul, 45, 63, 114, 156, 167–9, 171
Davis, S.T., 170
Dawkins, Richard, 78–9, 82–4, 86–8, 156
Dead Sea Scrolls
Deduction, 8, 10
Defective design, 101
Deism, 47
Demons, 128
Descartes, 1, 35
Designer, 59, 62, 156
Dharma, 48
Dinosaurs, 95
Dirac, Paul, 27
Disorder, 93
Divine experimenter, 45
Divine mathematician, 157
Divine scientist, 69
DNA, 21, 71, 77–81, 101, 118, 156–7
Dodd, C.H., 150
Dover, Gabriel, 83, 168
Dualism, 114, 136
Dual Model of God, 121
Dyson, Freeman, 134

Ecclesiastes, 101
Eigen, 85
Einstein, Albert, 2, 10, 14, 16–18, 30, 62, 67, 91, 115
Election, 88
Electron, 16, 41
Eli, 159
Elijah, 129
Eliot, T.S., 108, 133
Elisha, 34, 143
Empedocles, 76
Emergence, 115
Empirical, 87
Enlightenment, 142
Enoch, 138
Esau, 52
Eschatology, the debate, 150–3
Evidence, 83
Evil, ch. 6
Evil impulse, 69
Evolution, 75–7, 92, 99, 159–60

Faith, 3, 34–5
Fall, 86–7, 98
Fallen angels, 94
Falsification, 10, 35
Father, 45
Fenwick, Peter, 135
Feuerbach, 55
Feyerabend, Paul, 15
Flood, 51
Forms, 68
Fossil record, 76
Fourth Gospel, 30, 46, 68, 123, 163
Free will, 90–1, 103, 161
Freud, 10, 55, 88
Fuller, M., 165, 170
Fuller, R.H., 146–7, 169–70

Galileo, 8, 14, 19, 39, 44, 63
Galloway, A.D., 170

Gambler God, 90
Gamlin, L., 168
Genes, 78–81, 83, 135
Genesis, 67
Genetic drift, 85
Genetic engineering, 80
Gnosticism, 35
God as father, 158–9
God as grand chess-master, 73
God as king, 47
God and metaphor, 40
God as mother, 158
God as personal, 47
God as scientist, passim
God of hope, 142
Gödel, Kurt, 17
Gospel records, 128–30
Gravity, 134
Grayson, Brian, 135
Gribbin, J. and M., 165–8

Habakkuk, 108
Hades, 142
Hadrons, 61
Haldane J., 19
Handel, 28
Hanlon, M., 170
Hardy, Alister, 81
Hartshorne, Charles, 70, 121, 149
Hawking, Jane, 104, 165, 169
Hawking, Stephen, 11, 19, 57–8, 64, 157
Heaven, 142
Heisenberg, Werner, 16, 18, 35
Hell, 142
Hertz, Heinrich, 14
Hidden God, 102
Hick, John, 49, 50, 55, 88–9, 94–5, 98–9, 142, 152, 166, 168–9, 170
Hindu, 24
Hinduism, 27, 48, 139
Holmes, Sherlock, 7
Holocaust, 101–4, 138
Holy Spirit, 32, 141–2
Horowitz, Sara, 169
Hosea, 34
Houghton, J., 169
Hubble, 60
Hume, David, 8, 63, 122–3, 155
Huyghens, 27
Huyssteen, J.W. van, 165, 167–8
Hyperbole, 51

Ignatius, 146
Image of God, 66, 87, 153
Imams, 28
Immortality, 137–8
Immutability of God, 71
Impassibility of God, 103
Incarnation, 47, 130
Induction, 7–8
Inflation theory, 60
Information, 112–13
Information processing, 156
Information processors, 134
Invisible entities, 19, 24, 26
Irenaeus, 88, 98–9
Isaiah, 40, 94
Islam, 48
Israel, 69, 102

Jacob, 51, 95
James, 109
Jeans, Sir James, 62
Jeremias, 150
Jesus, 31–9, 45, 49–51, 54, 68, 89, 95–7, 106, 109, 141, 160
Job, 45, 104–5, 159–60
John the Baptist, 26
Johnsson, J., 166
Jonah, 28
Joseph, 26, 89, 95, 103
Judaeo/Christian tradition, 5
Judaism, 27, 137
Judas, 73
Jung, 88
Jupiter, 59

Kant, Immanuel, 31, 900, 104, 118, 136, 153
Karma, 105–6, 140
Kekule, A., 25
Kendall, D., 170

Kenosis, 69–72, 158
Kepler, Johann, 14, 44, 63
Kingdom of God, 41, 129, 131
Klotzko, A.J., 170
Kohler, K. 169
Kolitz, S., 169
Krishna, 96
Kuhn, Thomas, 11–12, 20, 43–4, 165

Laboratory of the world, 96
Lakatos, Imre, 12, 19
Lamb, 38–9, 90
Language, 3
Lavoisier, 107
Laws of inheritance, 9
Laws of nature, 123–4
Lazarus, 126
Leah, 51
Leaven, 131
Lemaitre, Georges, 18
Liberation theology, 33
Life after death, 148–9
Limitations of science, 15ff.
Linear time, 137
Lion of Judah, 39
Lister, 107
Literal, 38
Literary devices, 51ff.
Loaves and fishes, 125
Logos, 46, 60, 67, 157, 162
Lourdes, 123
Lovelock, James, 15
Luscombe, P., 165
Luther, Martin, 24–5

Maccabean, 138
MacGrath, A., 166
Magdalene, Mary, 145
Mahayana Buddhism, 48, 139
Maitreya, 47
Many universes, 59
Mars, 59
Mars Hill, 163
Martyr, Justin, 146
Materialist, 89
Mathematical models, 43
Matter, 16
Matthews, R., 170
Marx, Karl, 10
Maxwell, James Clerk, 15, 23, 27
McMullin, Ernan, 42
Memes, 76, 84
Mendel, Gregor, 9
Mercury, 59
Messiah, 33, 137–8
Metaphor, 39–41
Michal, 51
Migliore, D., 166
Milton, 98
Mind, 113
Mind behind the universe, 162
Mind/brain, 114–19
Miracles, 122–31
Models, 41–2
Models of God, 43
Molecular drive, 85
Moody, Raymond, 135
Moltmann, Jurgen, 71, 167
Morality, 153
Moral sense, 91
Moses, 26, 40, 68, 129
Mother, 46
Muhammad, 24, 26–7, 33, 48, 54, 139
Multiple attestation, 129
Murphy, Nancey, 25
Muslim, 138
Mutations, 80
Myth, 49
Myths of creation, 65

Nanak, Guru, 24, 48, 54
Natural selection, 42, 80, 82
Neanderthals, 137
Near-death experiences, 135
Necessity, 63
Neo-Darwinism, 76–7, 80
Neptune, 9
Newton, Isaac, 7–10, 12, 14, 18, 27, 61, 63, 91, 107
Nineveh, 89
Nirvana, 140–1
Noah, 69

Oakley, Susan, 145
Objections to model of God, ch. 9
Observation, 7–8, 10
Observer, 2–3
O'Collins, G., 170
Omnipotence, 25, 72–3, 93–4
Omniscience of God, 72–3
Oppenheimer, Robert, 108
Organisms, 75–6, 83
Origen, 51
Origin of evil, 94–5
Orthodox Judaism, 29

Padmanabhan, Thanu, 167
Palestinians, 144
Panentheism, 111, 121
Pannenberg, W., 28, 144, 150–2, 170
Paradigms, 11, 24–5, 44
Paradise, 149
Paradoxes, 16, 41
Parthenon, 163
Passover, 38
Pasteur, 76
Paul, 39, 46, 51, 86–7, 92, 94, 97, 108–9, 145–6, 159, 162
Peacocke, A.R., 70–1, 85, 111–14, 127, 167–8
Penrose, Roger, 44–5, 58, 118, 136, 166, 169
Pentateuch, 29, 51
Personal, 47, 158
Persons, 116, 162
Peter, 96
Phillips, D.Z., 52–5, 167
Philo, 67, 157
Planck, Max, 27
Plantinga, A., 95, 168
Plato, 31, 50, 67, 114, 157
Polanyi, Michael, 14, 19, 29, 166
Polkinghorne, John, 27, 71, 111–14, 127, 142, 148–9, 169–70
Poole, M., 64, 165–7
Popper, Karl, 10–11, 13, 53, 85
Postmodern, 142
Power of God, 158
Pre-existence of Christ, 69
Prigogine, I., 85
Principle of simplicity, 14
Problem-solving, 10, 32
Process theology, 70, 119–21
Prodigal Son, 38
Psalmist, 2, 96–7, 138, 158
Psychic phenomena, 127
Psychologists, 3
Punctuated equilibrium, 76
Purgatory, 143, 149
Purpose of God, 54, 81
Purposeful humanity, 84

Quantum effects, 57
Quantum entanglement, 161–2
Quantum particles, 136
Quantum physics, 8, 15–17, 21, 64
Quarks, 24, 61
Qur'an, 26–7, 38, 48

Rakover, Yossel, 102
Reconstructionists, 29
Recycling, 133
Red Sea, 126
Reform Judaism, 29
Religious experience, 24, 33
Religious language, Ch. 3
Remnant, 69
Replicas, 134
Resonance level, 59
Resurrection, 71; of Christ, 125, 143–8, 151, 163
Revelation, 26–9, 31–2, 34
Revolution, 11
Reynolds, Pam, 136
Richardson, Alan, 51, 166
Rolston, Holmes, 71
Rome, 141
Rose, Steve, 119, 169
Russell, Bertrand, 7
Ryle, Gilbert, 119

Sabbath, 68, 130, 147
Sadducees, 138
Saga, 51
Samsara, 105
Sanders, P., 148
Sankara, 121

Satan, 94, 96
Saul, King, 89
Saul of Tarsus, 34
Schweitzer, Albert, 150, 152
Scott, A.C., 167
Scripture, 25, 28–30,
Self-conscious, 87, 90
Self-imposed limitations of God, 72
Serpent, 66, 94
Sheol, 142, 149
Shi'ah, 28, 48
Shiva, 85
Signs, 123
Sikhism, 24, 48, 98
Singularities, 31, 58
Skandhas, 140
Snedden, R., 168
Social sciences, 2, 158
Social scientist, 158
Sociobiology, 79
Sociologists, 3, 20
Socrates, 8
Son of God, 50, 129, 151
Soul, 66, 91, 137, 139–40, 153
Species, 39
Spinney, L., 168
Stannard, Russell, 88, 109, 167–9
Stein, R.H., 166
Stewart, I., 167
Strawson, Peter, 116
Subordination, 72
Suffering, 99–100, 102, 159–60, Ch. 6
Suffering God, 72–3, 103
Suffering scientist, 106–8
Suffering servant, 106
Sufism, 48
Sun, 133–4
Sunni, 28
Superstring theory, 60–2
Survival of the fittest, 99, 159
Swinburne, Richard, 52, 145
Symbols, 38, 41, 67

Targums, 67
Temim, Howard, 81
Temptation, 96–7
Teresa, Mother, 86
Testing, 14f, 34, Ch. 6
Theories, 10, 12, 15
Theory, 3, 33
Tillich, Paul, 38
Time, 16, 62
Tipler, Frank, 134–5, 170
Top-down, 113
Torah, 51
Torrance, Tom, 29–31, 166
Tower of Babel, 51
Travis, S.H., 170

Uncertainty principle, 16
Universe as an organism, 63
Upanishads, 139

Values, 47
Vanstone, W.H., 70
Vardy, Peter, 82, 93, 167–9
Various models of God, 155
Veda, 139
Venus, 59
Virgin birth, 126
Vishnu, 27, 47, 96

Walker, Alan, 108
Walking on the sea, 127
Wallace, A.R., 77
Ward, Keith, 45, 64, 91–2, 111, 126–7, 149, 166, 168–70
Washoe, 117
Watson, W., 78
Weatherhead, Leslie, 167–9
Weiss, 150
White, M., 167–8
White, S., 170
Whitehead, A.N., 120–1
Wiles, Maurice, 112, 123
Wilson, E.O., 119, 169
Wisdom, 67, 138
Witness of miracle, 125
Wittgenstein, 52–3
Word, 67–9
Wright, R.T., 165

Yahweh, 69, 150
Yoga, 139
Zaehner, R.C., 170
Zarate, O., 169
Zygmuntowicz, Itka, 104, 169